SCHWATKA'S
LAST SEARCH

T0136771

SCHWATKA'S

LAST SEARCH

THE NEW YORK LEDGER EXPEDITION
THROUGH UNKNOWN ALASKA
AND BRITISH AMERICA

INCLUDING THE JOURNAL OF CHARLES WILLARD HAYES, 1891

WITH AN INTRODUCTION
AND ANNOTATION BY

ARLAND S. HARRIS

University of Alaska Press
Fairbanks

Library of Congress Cataloging-in-Publication Data

Schwatka, Frederick, 1849–1892.
  Schwatka's last search : New-York ledger expedition through
unknown Alaska and British America : including the Journal of
Charles Willard Hayes, 1891 / with an introduction and annotation by
Arland S. Harris.
    p.   cm.
  Schwatka's narrative of his 1891 exploration in Alaska published
in the New-York ledger magazine as a series of installments in 1892.
  ISBN 0-912006-87-0 (pbk. :alk. paper)
  1. Alaska--Discovery and exploration--American. 2. British
Columbia---discovery and exploration--American. 3. Yukon Territory--
-Discovery and exploration--American. 4. Schwatka, Frederick,
1849–1892--Diaries. 5. Explorers--United States--Diaries.
6. Hayes, C. W. (Charles Willard). 1859–1916 Diaries.   I. Hayes,
C. W. (Charles Willard). 1859–1916. Journal of Charles Willard
Hayes, 1891. II. Harris, Arland S. III. New-York ledger.
IV. Title.
F908.S39   1996
979.8'01--DC20   96-38514
                                                       CIP

International Standard Book Number: 0-912006-87-0
Library of Congress Catalogue Number: 96-38514

Printed in the United States by BookCrafters, Inc.

This publication was printed on acid-free paper that meets the minimum
requirements of American National Standards for Information Sciences-
Permanence of Paper for Printed Library Materials, ANSI Z39.48-1984.

Publication coordination and production by Deborah Van Stone, University
  of Alaska Press.
Publication and cover design by Dan Kaduce and Dixon Jones, Rasmuson
  Library Graphics, University of Alaska Fairbanks.

Contents

LIST OF ILLUSTRATIONS

LIST OF MAPS

FOREWORD

For the historian, finding a significant, supposedly lost document is like the excitement an explorer experiences in viewing a new valley, pass, river, or mountain. Arland Harris had it both ways, in a manner of speaking. He located the account by Frederick Schwatka of his last and most important exploration in Alaska: his 1891 trek through the headwaters of the White River, the Skolai Pass, and the upper Chitina drainage, which comprised one of the last big chunks of the country to feel the boot of Euro-American scientist-explorers. At the same time, Harris enjoyed the thrill of knowing that he was probably the first student of Alaskan exploration, in this century, to experience vicariously how Schwatka himself reacted to the unfolding landscapes theretofore unreported in the annals of western geographical discovery. I know the feeling, but not because I found any lost documents, only documents that had been ignored by most historians; and—alas!—I was partly responsible for losing the narrative that Harris found in *The New York Ledger* magazine.

Actually, I did not lose it; I never found it, and hypothesized its nonexistence, as did that skilled and conscientious writer, Eliza Ruhamah Scidmore.[1] There are good reasons for the bad guess. Schwatka had the temperament (and talent) of a popular writer and lecturer which is what he became after quitting the army in 1885, two years after he rafted the Yukon River on his most famous geographical reconnaissance. He published accounts of that feat before his official report was issued, and a commercial book about the trip appeared in the same year as the

---

1. Sherwood, M. *Exploration of Alaska, 1865–1900* (New Haven and London: Yale University Press, 1965; reprinted Fairbanks: University of Alaska Press, 1992) 143. Scidmore, E. R., "Recent Explorations in Alaska," *National Geographic Magazine*, 5 (January 31, 1894) 176.

federal narrative. (His boss, General Nelson Miles, was not pleased with Schwatka's rush into print.) There was reason to believe that he would publish soon after his 1891 trip too, but he died within months of his return from Alaska. Judge James Wickersham's bibliography of Alaskan literature covering the years to 1924 made no mention of *The New York Ledger* series; some of us had come to rely on the thoroughness of Wickersham (or rather, Hugh Morrison of the Library of Congress, who did most of the work). No other guide or related source mentioned a Schwatka journal. In scanning the newspapers available in Juneau I missed seeing the brief notice in the *Sitka Alaskan* for May 30, 1981, mentioning *The New York Ledger.* I missed the *Juneau City Mining Record* completely; in extenuation, it may not have been available and certainly was not on microfilm at the time I did my research in the early 1960s. Furthermore, Schwatka was one of only two national celebrities among the many accomplished explorers of Alaska between 1865 and 1900.[2] How could this most important expedition, if published, escape notice, especially when Schwatka's companion, the geologist C. Willard Hayes, said afterwards that a "syndicate of newspapers" sponsored the journey? So, I adopted Scidmore's reasonable conclusion that no journal had been published. I recount this thinly disguised argument to excuse my own errant scholarship, but also to illuminate how historical research proceeds—often tediously and awkwardly yet often like a detective's investigation with the excitement of original geographical discovery.

Now, Arland Harris has established that Schwatka did hurry into print when desperately ill, and Harris has edited the dramatic story of Schwatka's 1891 Alaskan exploration, his most significant when measured by the difficulty of the terrain and the original contributions to geographical knowledge. In the book that follows, and in personal communications, Harris has resurrected Schwatka's lively narrative and repaired his reputation as well. On occasion, Schwatka has been handled roughly by contemporaries and historians. George Dawson, the celebrated Canadian geologist, and William Healey Dall, the great American natural historian, both scolded Schwatka for his liberal use of new toponyms and for his disparagement of previous cartography, while

---

2. Sherwood, M. "American Explorers of Alaska," in A. Shalkop, ed., *Exploration in Alaska,* (Anchorage: Cook Inlet Historical Society, 1980) 147.

they admitted that his map of the Yukon River, made in 1883, was
very good. Alfred Hulse Brooks credited Private Charles Homan with
the map, not Schwatka. Historian Melody Webb thinks his toponymy
represented "arrogant disdain for Canadian sensibilities...." Scidmore
said the Yukon River voyage was "not discovery in any sense...." I
characterized the natural science that came from his St. Elias expedition of 1886 as insignificant, an English geographer thought that any
attempt to guide Schwatka "of eighteen stone" up Mount St. Elias was a
"Quixotic enterprise," and a New York newspaper satirized in rhyme
the name Schwatka gave to an Alaskan river. Then his reputation as a
resolute cold-weather explorer suffered in 1887 when, on a planned
winter trek through Yellowstone Park, he collapsed after four days and
twenty-six miles. In my attempt to do an aggregate biographical study
of Alaskan explorers, I called Schwatka a "promoter," which he was, but
the word connotates unacceptable qualities. (That he engaged in a
Mexican land and colonization scheme with Buffalo Bill Cody probably
did not help Schwatka's reputation for modesty.) The noted historian of
Alaska, Ted C. Hinckley, saw in Schwatka's West Point photograph a
"devil-may-care swagger." Allen Wright calls him "a dapper, arrogant,
little man," though the "little man" was well over six feet in height and
weighed one-eighth of a ton.[3]

Editor Harris interprets the explorer's character and personality rather
differently. He points out that Schwatka had a sense of humor and a
fondness for practical jokes at West Point, sometimes at his own expense, and that his humor is revealed in his writings. He was a convivial
person and adventuresome. He elected to leave the army and be a full-
time explorer, writer, and lecturer, which required self-advertisement
and promotion to succeed; it was a "business necessity," contends
Harris. He is right, and at least Schwatka was not a pseudo-celebrity in

3. Sherwood, M. *Exploration of Alaska,* 77–79, 101, 102. Brooks, A. H. *Blazing Alaska's Trails* (Fairbanks: University of Alaska Press, 1953) 275. Webb, M. *The Last Frontier: A History of the Yukon Basin...* (Albuquerque, New Mex.: University of New Mexico Press, 1985) 106. Scidmore, "Recent Explorations," 174. Lang, W. L. "'At the Greatest Peril to the Photography...'" [in Yellowstone} *Montana,* 33 (winter 1983) 14–25. Sherwood, "American Explorers...," 146. Russell, Don, *Lives and Legends of Buffalo Bill* (Norman, Okla · University of Oklahoma Press, 1960) 423, 424. Hinckley, "Review," *Journal of the West,* 4 (April 1965) 278. Allen A. Wright, *Prelude to Bonanza* (Sidney, British Columbia: Gray's 1976) 142.

the sense that Daniel Boorstin decries.[4] Schwatka actually did something worth admiring and advertising. He did scatter new geographical place names over the land too eagerly but many of those names remain for reasons of priority and because sometimes he honored men of science. (Other explorers have succumbed to the same temptation. Even Dall named a small island in Kachemak Bay, southcentral Alaska, after a British aristocrat whose small yachting party's main activity was to blast away happily at the country's wildlife, of all species and sizes, in one of the most extravagant animal slaughters seen in a territory that has seen many.)[5] Though Schwatka collapsed when on his St. Elias expedition and in Yellowstone, and was in serious physical pain on the 1891 journey, he was a person of immense physical strength and energy, as the successful expeditions demonstrate clearly. None of the failures prevented him from planning new tests of his mettle. He possessed intellectual energy, too. During a mere handful of years he managed to study law and be admitted to the Nebraska bar and to study medicine and receive a doctorate, all while he was fighting the Sioux or completing a three thousand-mile arctic sledge journey. A medical doctor writing in a technical journal thinks that Schwatka "should be remembered as a pioneer American environmental physiologist" and expert on arctic-mountain survival.[6]

One thing Frederick Schwatka was, most assuredly: a skilled storyteller and literary craftsman in the exploration genre so popular during the second half of the nineteenth century. His narrative provides a robust counterpoint to scientist Hayes's factual, sober but more instructive account of the same journey through uncharted wilderness previously seen only by Native Americans.

                                                            —Morgan Sherwood

---

4. Boorstin, D. *The Image: A Guide to Pseudo-Events in America* (New York: Harper's, 1964).

5. Francis, F. *War, Waves, and Wanderings: A Cruise in the "Lancashire Witch"* (London: Sampson et al. 1881; two vols.) II, 218–287 for the Alaska trip. The aristocrat was Sir Thomas Hesketh.

6. Johnson, R. E. "Medical Intelligence: Doctors Afield—Frederich Schwatka…." *New England Journal of Medicine,* 278 (January 4, 1968) 34.

PREFACE AND ACKNOWLEDGMENTS

The early exploration of Alaska and northwest Canada has intrigued me since I first arrived in Alaska over forty years ago. Since then, thirty years of field work in forest research in coastal Alaska and excursions over the years in interior Alaska, northern British Columbia, and Yukon have allowed me to see where some of this exploration took place.

I'm not sure when I first read Morgan Sherwood's book, *Exploration of Alaska 1865–1900*, first published in 1965, but I found that it was a fine introduction to the decades of exploration preceding the great Klondike gold rush. It was also my introduction to the explorer Frederick Schwatka.

Many years later when I decided to learn more about Schwatka's little-known expedition of 1891 through Alaska, British Columbia, and Yukon, Professor Sherwood's well-documented pages helped to direct my search. At the time I had no idea that I would later owe him thanks for his review and encouragement of my effort to document the journey.

In 1892 Charles Willard Hayes, who accompanied Schwatka, wrote in the *National Geographic Magazine* that the expedition was sponsored by a syndicate of newspapers and that Schwatka's personal narrative would appear elsewhere. But historians agreed that Schwatka died before his narrative could be published.

I began my research with the idea that even if Schwatka's narrative had not been published, such a famous explorer would surely have kept a journal or notes and that these might be located. My search proved fruitless. Then I reviewed contemporary Alaskan newspapers and found the answer—articles in the Juneau and Sitka newspapers of 1891 named a single sponsor of the expedition: *The New York Ledger.*

Further searching revealed that Schwatka's narrative had indeed been published in *The New York Ledger* magazine as a series of installments in 1892. By 1990 I had a copy of Schwatka's narrative in hand. Here was the document that had eluded historians for a century. Finding it was a thrill I will always remember.

With the help of the U.S. Geological Survey (USGS) I learned of and obtained copies of Hayes' journal and photographs made during the expedition. Since then, in a long off-and-on effort I have tried to understand the details of the expedition and to learn something about the lives and personalities of the participants.

Until recently I thought that I was alone in rediscovering Schwatka's narrative. Then I found Allison Mitchum's book, *Taku: The Heart of North America's Last Great Wilderness* (1993) and learned that Professor Mitchum had also discovered Schwatka's narrative and had reprinted excerpts from the portion of his journey along the Taku River and in northern British Columbia.

There is much of historical, geographical, and general scientific interest in Schwatka's and Hayes' accounts. Together they offer a glimpse of what exploration was like on the northern frontier as described by two very different men: one an observant though flamboyant explorer, writer, and lecturer, writing to entertain the public of his day; the other a well-educated scientist and geologist fascinated with the natural world. In presenting the two accounts I have tried to intrude as little as possible and let each man tell his own story.

Over the years many people and institutions have helped in my search and I am grateful to all. I thank:

The staff of the Alaska State Library, Juneau, Alaska for making me fell welcome and for providing continuing help, especially Kay Shelton, India M. Spartz, and Gladi Kulp of the Alaska Historical Collections and Ron Reed, Jeff Rothal, and Elaine Hobbs of Interlibrary Loan.

Elizabeth Behrendt, Technical Information Specialist, USGS Earth Science Information Center, Anchorage, Alaska, for help in locating Schwatka's narrative and Hayes' journal.

Jill L. Schneider, Geologist, USGS Technical Data Center, Branch of Alaskan Geology, Anchorage, Alaska, for providing the copies of Hayes' journal.

Isabella Hopkins and Joe McGregor, USGS Photo Library, Denver, Colorado, for providing photographic prints and negatives.

Jeffrey H. Kaimowitz, Curator, Watkinson Library, Trinity College, Hartford, Connecticut for providing a microfilm copy of the 1892 issues of *The New York Ledger.*

Dr. Robert E. Johnson, M.D., South Burlington, Vermont for providing much biographical information about Schwatka and for encouragement in my search.

Ron Klein, Northlight Studio, Juneau, Alaska for help with the photographs.

I also greatly appreciated the help provided by the following people and organizations in locating information:

Jane Albrecht, Library Technician, U.S. Bureau of Mines, Library, Juneau, Alaska.

Annette Bartholomae, Reference Librarian, Oregon Historical Society, Portland, Oregon.

Judy Belan, Special Collections Librarian, Augustana College, Rock Island, Illinois.

Virgina Deffner and Katy Powers, Reference Department, Rock Island Public Library, Rock Island, Illinois.

Thomas A. Dietz, Curator of Collections, Kalamazoo Public Museum, Kalamazoo, Michigan.

Richard Fusick, Archivist, U.S. National Archives, Civil Reference Branch, Washington, D.C.

Jean Gillmer, Northwest Room, Tacoma Public Library, Tacoma, Washington.

Joan Grattan, Manuscripts Assistant, Milton S. Eisenhower Library, The Johns Hopkins University, Baltimore, Maryland.

Terry Harrison, Historian/Archives and Robert Bowen, Research Volunteer, Mason City Public Library, Mason City, Iowa.

Carol S. Jacobs, Archival Assistant, Oberlin College Archive, Oberling, Ohio.

Alfred C. Jones, Marion County Historical Society, Salem, Oregon.

James V. T. McEnery, Special Collections Division, U.S. Military Academy Library, West Point, New York.

Marjorie Napper, Oregon State Library, Salem, Oregon.

James C. Roberts, Reference and Bibliography Section, U.S. Library of Congress, Washington, DC.

Barbara Scott, Rock Island County Historical Society, Rock Island, Illinois.

Richard A. von Doenhoff, U.S. National Archives, Military Reference Branch, Washington, DC.

Special thanks also to Debbie Van Stone and the University of Alaska Press for much needed help in editing and publication.

My most grateful appreciation goes to my wife, Bina, a cheerful companion on portions of Schwatka's travels, for help in writing, for essential computer aid and especially for putting up with my long obsession with Frederick Schwatka and the New York Ledger Expedition.

INTRODUCTION

I n 1897 news of the Klondike gold strike electrified the world. The new Eldorado lay far to the north in a land almost unknown to the public, but within the year stampeders could buy maps and guide books describing the Yukon basin, gold fields, trail routes and passes through the massive coastal mountains. Alaska and the adjacent Canadian Interior were not so unknown after all.

Native groups had known their own territories within this region for centuries, and had passed their knowledge on to the Russians, to British and American traders, and to missionaries who came early in the 19th century. Later in the century scientific and systematic mapping began to reveal the full extent of the land.

The two decades preceding the Klondike strike brought unprecedented exploration in Alaska and northwest Canada, now British Columbia and Yukon. Expeditions were sponsored by governments, scientific societies, and newspapers. Explorers included military officers, government agents, journalists, and adventurers. Their motives varied, but they shared a great curiosity and were willing to endure hardship and to risk unknown dangers.

Frederick Schwatka was the most celebrated late-nineteenth-century explorer to leave his mark on the maps of Alaska, Yukon, and northern British Columbia, and in the course of three expeditions he named more physical features than perhaps any of his contemporaries. Travelers in the region today can hardly avoid coming across his tracks.

Schwatka's expedition of 1891 was his most important, yet it is little remembered, and his narrative has been lost to modern readers. Schwatka made his living from exploring, writing, and lecturing. Aside from indulging his love of exploration, his main object in making the expedition was to cover as much unmapped country as possible, thereby

1

Frederick Schwatka's class portrait, West Point, Class of 1871.
(U.S. Dept. of the Army, West Point Military Academy Library)

adding to his reputation as an explorer and providing lively copy for his sponsor, *The New York Ledger.* His personal narrative is reproduced here.

The *Ledger*, owned and published by the Robert Bonner family, billed itself as "A journal of choice literature, romance and useful information." It was always in need of lively copy by prominent writers, and the editors were glad to feature Schwatka's vivid writing. With a lifetime of colorful experience to draw on and his knack for embellishing a story, he was a favorite with the public, and his friendly, jovial manner assured his popularity on the lecture circuit.

Frederick Schwatka was born at Galena, Illinois, in 1849, and four years later came west with his family by ox-team over the Oregon Trail.[1] The family homesteaded near Astoria, then moved to Salem in 1859, where Frederick attended public schools and studied briefly at Willamette University before working as a printer. In 1867 he received an appointment to West Point, graduating four years later with a commission as second lieutenant. He was only an average scholar, but was well thought of by his classmates at West Point who appreciated his sense of humor. One later wrote:

> There was no state of affairs, however distressing, in which he was not able to discover a ridiculous side, or out of which he was not able to deduce a humorous conclusion. He was always ready for a practical joke, in which he was quite willing to be accounted the principal sufferer, so long as it contributed to the general stock of amusement. A harmless irony and a quaint humor adorned his conversation, and made him always a welcome visitor in the rooms of his classmates and friends.[2]

On graduation Schwatka was assigned to the Third Cavalry at Fort Apache, Arizona. He soon transferred to the Department of the Platte and for the next five years served at duty stations on the northern plains and in campaigns against the Sioux led by Brigadier General George Crook—the Yellowstone Expedition in 1872 and the Big Horn and Yellowstone Expedition in 1876, where he saw action at Tongue River, The Rosebud, and Slim Buttes.[3]

Always ambitious, Schwatka studied law and medicine on his own time, as circumstances allowed. Self-study and reading in chambers with lawyers accounted for much of his legal preparation, and by passing the required examination he was admitted to the bar for the District of

1. Lieutenant Schwatka's father, Frederick G. Schwatka, was born in Maryland, of Polish descent. He married Amelia Hukill of West Virginia in 1833. Frederick was the seventh of eight children (Johnson 1984, p. 3). (*See also* Hodgkin and Galvin 1882, p. 65.)

2. Association of West Point Graduates Annual Report for 1893, pp. 67–68.

3. 47th Congress, 1st Session, H.R. Report No. 933, April 6, 1882. Schwatka's part in the battle of Slim Buttes was described by Finerty (1961, pp. 187–188).

Nebraska in May of 1875. His fellow lawyers later remembered him as intelligent, congenial, convivial, and at times a hard drinker.[4]

He was also interested in medicine. At the time, an apprenticeship, a short residence at a medical college, and passing an examination qualified a candidate as a medical doctor. Schwatka had assisted army surgeons in Arizona and while campaigning against the Sioux, and over the years had managed to spend some of his leave time in residence at Bellevue Hospital in New York. He passed the examination, wrote a thesis on diphtheria, and received a medical degree from Bellevue Medical College in New York City in 1876.[5]

During the next two years he served as adjutant at the Spotted Tail Agency in northwestern Nebraska where his friendship with the Sioux chiefs, Spotted Tail and Standing Elk, led to his adoption into the tribe under the name "Great Wolf."[6]

Schwatka was an avid reader of arctic exploration, and since his youth he had been fascinated by the fate of the lost Franklin Expedition. Since the Franklin party's disappearance in the Arctic in 1845 while attempting the Northwest Passage in the ships *Erebus* and *Terror*, searchers had turned up artifacts and bones, but log books and scientific records remained lost. When Schwatka learned that yet another search was being organized by the American Geographical Society, he offered his services. Judge Daly, President of the society, invited him to New York and on the basis of an interview offered him the command of the expedition. Daly's request to Secretary of War General William T. Sherman that Schwatka be allowed leave of absence from the army was granted and he was placed on detached service at half pay.[7]

The expedition, with the newspaperman William Henry Gilder as second in command, sailed from New York on June 19, 1878, in the schooner *Eothen* and returned on September 22, 1880. During their search on King William Island in the Canadian Arctic they were away from their base of supplies for eleven months and twenty-three days, living off the country with the Inuit and traveling by dog sled on a reported journey of 3,251 miles—the longest sled journey then on record.

---

4. Johnson et al. 1984, p. 6.

5. Johnson 1968, pp. 31–32.

6. Schwatka's vivid article in *The Century Magazine* described the Sun Dance of the Sioux (Schwatka 1890).

7. 47th Congress, 1st Session, H.R. Report No. 933, April 6, 1882.

They brought back relics of the Franklin party, and found the grave of Lieutenant John Irving, third officer of the *Terror*, but failed to find any documents. But most important, they proved that white men could subsist in the Arctic by adopting Eskimo ways. This meant, among other things, subsisting on a diet of raw meat and fat. Despite this diet Gilder described Schwatka as remaining in robust health, "with a powerful frame to which fatigue seems a stranger," adding, "...he possesses a very important adjunct, a stomach that can relish and digest fat." Gilder's story of the expedition in the *New York Herald* and his book, *Schwatka's Search* (1881), caught the public's fancy and made Schwatka an instant celebrity. "Schwatka's Search" became a popular phrase denoting the dogged pursuit of a goal in the face of heavy odds.[8] He returned to the United States a hero and received many honors including the Roquette Arctic medal from the Geographical Society of Paris and a medal from the Imperial Geographic Society of Russia.

On his return to active duty in 1881, Schwatka was appointed aide-de-camp to General Nelson A. Miles, then commanding the U.S. Army's Department of the Columbia, headquartered at Fort Vancouver, Washington Territory. On his way from New York to the west coast Schwatka took leave in Rock Island, Illinois, at the invitation of Colonel A. G. Brackett, formerly of the Third Cavalry, where he courted the Colonel's niece, Ada J. Brackett.[9] The following year Frederick and Ada were married, and the couple settled in at the married officer's quarters at Fort Vancouver.

Two years later Schwatka made his first Alaskan expedition. In 1883 the army lacked jurisdiction in Alaska, but General Miles was certain that in case of trouble with Alaska Natives in the Interior, the army would be called on to quell any disturbance. Maps of interior Alaska were inadequate and his efforts to get congressional approval for systematic mapping failed. But Miles was determined to get information on the country and the potential fighting ability of the Alaska Natives, and assigned Schwatka the job.

The most feasible route to Alaska's interior lay across Canadian territory, and obtaining official permission was a formality that General Miles wished to avoid. Schwatka organized a secret expedition, and

---

8. While on this arctic expedition, Schwatka was promoted to First Lieutenant.

9. The couple had first met earlier in the year while Ada was visiting her uncle in Wyoming (*New York Times*, September 11, 1882, p. 4).

with six companions quietly set off from Puget Sound by commercial steamer for Chilkat,[10] at the head of Lynn Canal, near the present city of Haines. Here he hired Native packers, and as he later wrote, the party left "like a thief in the night," taking the well-known "miner's route," up Taiya Inlet, Dyea River, and over Chilkoot Pass (which Schwatka named Perrier Pass), to Lake Lindeman.[11] Here they built a raft and continued through Lake Bennett—named by Schwatka to honor James Gordon Bennett, owner-publisher of the *New York Herald*, and on down the Yukon River to its mouth. Along the way they visited Native villages, and although the inhabitants were in all cases friendly, Schwatka noted carefully their potential defenses.[12]

When Schwatka completed his tour of duty under General Miles, he rejoined the Third Cavalry in Arizona. By 1885 the Indian wars were nearly over and the army was being reduced in size. Schwatka could see little chance for promotion, and so resigned his commission, much to his wife Ada's distress. Henceforth he intended to continue exploring and to support his family through lecturing and writing.

In 1886 Schwatka made a second expedition to Alaska, this time an attempted first ascent of Mount Saint Elias, the 18,000-foot peak on the Alaska-Yukon border near Yakutat. The expedition was sponsored by the *New York Times*, and included Dr. William Libbey, professor of geology at Princeton University, John Dalton,[13] roustabout and camp cook, and the English alpinist and travel writer, Heywood W. Seton-Karr. Schwatka and Ada traveled to Sitka by commercial steamer, where the expedition continued on in the U.S.S. *Pinta*, a Navy gunboat on station in Alaskan waters.

---

10. Chilkat was located on the eastern side of Pyramid Harbor at the mouth of the Chilkat River, and in 1891 was the most northerly port of call for steamships plying the Inside Passage. At the time of Schwatka's visit two salmon canneries and a trading store were located here (Moser 1899, pp. 16, 125).

11. Schwatka's name, "Perrier Pass," which he bestowed in 1883, never caught on and the Native name *Chilkoot* remains in use today. Lake Lindeman, which Schwatka named to honor the Secretary of the Bremen Geographical Society, has been retained.

12. Schwatka's *Report of a Military Reconnaissance in Alaska, Made in 1883*, was published by the United States Printing Office in 1885. The same year he published a popular account, *Along Alaska's Great River*, which was later reprinted in several editions (1893–1885) as *A Summer in Alaska*.

13. John (Jack) Dalton later built the Dalton Trail and operated it as a toll road between Chilkat at the head of Lynn Canal and the Yukon River (Coutts 1980, pp. 72–75).

**Frederick Schwatka on the lecture circuit dressed in Inuit parka.
The photo appeared in Schwatka's book, *Nimrod in the North*,
1885. (Author's collection)**

Although still a young man of thirty-six, Schwatka was much overweight and in poor physical condition. The privations endured in his
earlier years had taken their toll, as had the social life of the lecture
circuit. The climbing party reached a point twenty miles inland at an
elevation of about 5,700 feet when Schwatka collapsed, suffering from
symptoms described as extreme fever, chills, and pleuritic pain. The
expedition was forced to turn back after only nine days, far short of the
summit. Schwatka reported the trip in a series of articles in the *New
York Times*. He blamed the party's inability to proceed further on inaccessible terrain but failed to mention his collapse. He also reported his

discovery of a great river rivaling the Mississippi in size which he named, "Jones River," to honor Mr. George Jones, publisher of the *Times* and backer of the expedition. Competing New York papers ridiculed the name, maintaining that the grand river to which Schwatka had applied the prosaic name "Jones," was already known to earlier map makers and that it was really a comparatively insignificant stream. The *Times* tried to put the best face on events but it was obvious that the expedition had fallen short of success—to the embarrassment of the *Times* and the glee of its competitors.[14]

The expedition was further marred by the tragic death of three Natives who died after eating bread made with what they thought was baking powder, but later was found to be arsenic intended for preserving hides. The poison had been placed in an empty baking-powder tin and carelessly dropped or discarded by Dalton.[15]

The expedition was successful in that it added some useful geographical information. But perhaps its greatest contribution was a collection of Native artifacts from Yakutat Bay. This collection—including a major find taken from a shaman's grave—was added to Princeton's ethnographic collection by Professor Libbey.[16] Seton-Karr described the expedition in his book, *Shores and Alps of Alaska* (1887b), and Schwatka followed with an article later published in *The Century Magazine* (1891).

The following year (1887) Schwatka joined an expedition financed by the Northern Pacific Railroad and the *New York World* to attempt a first winter crossing of Yellowstone Park. The party consisted of eleven men, with F. J. Haynes as photographer. They had been on the trail only a short time when Schwatka collapsed, unable to continue. He was

14. Schwatka's narrative was published serially in the *New York Times* from June to November, 1886. (*See also* Sherwood 1992, pp. 78–79.) Schwatka first published the name "Jones" in the *Times* on September 20, 1886, and in *The Alpine Journal* in November (vol. 13, p. 90). The river is shown on a map in the *Times* on November 14 (p. 6). The river was known to the natives as the *Yahtse*, according to Williams (1889, p. 392), and a river of this name is still shown on modern maps. Glacial recession since Schwatka's visit has caused radical changes in the vicinity and Schwatka's original description of the river's size cannot be verified (see Orth 1967, p. 1063).

15. Seton-Karr 1887b, pp. 129–130.

16. Regarding the ethics involved, Baird (1965, p. 10) commented, "Since the line between grave-robbing and archaeology can be a very fine one, anthropologists rarely trouble their consciences about such deeds. In any case, the more perishable of the grave-goods had already been damaged by exposure and would certainly have deteriorated completely if they had not been salvaged by the mountaineering party."

forced to turn back, but Haynes and three others completed the trip. It was a bitter defeat for Schwatka, who worried that he had failed his backers, and especially humiliating because his father-in-law, Colonel Brackett, was a member of the party.[17]

Two years later Schwatka made a successful expedition to the Sierra Madre Mountains of Mexico, sponsored by the *America*, a Chicago newspaper, and Buffalo Bill Cody, with whom Schwatka planned land development.[18] He brought back the first well-documented ethnographic information on the Tarahumari Indians. He was still much overweight, by his own account tipping the scales at 267 pounds.[19]

The New York Ledger Expedition

By 1890 Schwatka's appearance hardly fit the popular conception of a famous explorer. Although well over six feet tall and powerfully built, he was still overweight, and his tiny pince-nez glasses gave him a decidedly owlish look. But what he may have lacked in athletic appearance he more than made up for in grit. Although he was still a popular writer and lecturer, Schwatka's reputation as an explorer had been tarnished by recent failures. He needed a successful venture. His thoughts again turned to Alaska and he set about organizing a third Alaskan expedition. In a letter to Major J. W. Powell, Director of the U.S. Geological Survey, Schwatka outlined his plans:

> ...I intend to conduct an exploring expedition through that portion of the unknown parts of Alaska bounded, roughly speaking, by the Yukon and White Rivers on the east, the Copper or Atna River on the west and the headwaters of the Tanana River to the northward.
>
> I believe that this unexplored part of our country will be found of more than usual interest in a geological sense, and probably in economic phases that would lead to further developments of the country. It would connect the geographical and geological surveys of the great

---

17. For an excellent account of the expedition, see Lang 1983.

18. Schwatka had fought with William Cody during the Sioux campaign when Cody was a scout with the U.S. Third Cavalry. Schwatka, Cody, and Dr. Frank Powell, a purveyor of patent medicine, later joined in an unsuccessful project to colonize 2.5 million acres of land in Mexico (Hebberd, 1952, p. 308).

19. Schwatka 1892a, p. 274. Schwatka further described the trip in a book, edited after his death by his wife Ada (Schwatka 1895).

Yukon River with those of the Pacific Coast via Copper river, and would be the only connection so made from a point on the length of that long stream.[20]

While Schwatka planned for the expedition, his health deteriorated. He was plagued by stomach problems, doubtless aggravated by his conviviality and the hard-drinking social life of the lecture circuit. He was in frequent pain and relied on self-prescribed doses of laudanum for relief.[21] On January 29, 1891, while on a lecture tour in Mason City, Iowa, he stumbled while climbing stairs and fell backwards over a bannister, landing on the floor below. He was knocked unconscious and sustained a broken nose and other facial injuries, wrenched his back and was thought to have internal injuries.[22] *The Mason City Times* commented:

> It is reported that he had been indulging somewhat heavily previous
> to the accident and had too much of a load on to make the landing at
> the head of the stairs. He is a talented man whose writings and lectures
> have a world-wide reputation and the good he has accomplished for
> science and history should go a long ways in curtailing harsh criticism
> incident to a bad but somewhat prevailing habit.[23]

The *New York Times* reported that he was not expected to live, and other newspapers reported his death. On learning the news Ada rushed to Mason City by train from their home in Rock Island, Illinois, to be with her husband. Schwatka's condition improved, and as soon as he could travel they returned to Rock Island. Apparently he was undaunted by the experience and continued planning for his expedition to Alaska.

To handle the scientific aspects of the expedition, Schwatka wrote to Major John Wesley Powell, director of the U.S. Geological Survey, formally requesting the services of a geologist. His request stated:

---

20. Schwatka to Powell, February 20, 1891. USGS. "Register to Letters Received, 1879–1901." NARC., Washington D.C., 20408. In his letter Schwatka faied to mention that the major portion of the proposed expedition would be within Canadian territory.

21. During the last years of his life Schwatka's physical problems apparently included obesity, alcoholism, and an undiagnosed stomach disorder.

22. *New York Times*, January 3, 1891, p. 3; *The Mason City Times*, Feb. 4, 1891, pp. 1–2, Mason City *Express Republican*, February 5, 1891.

23. *The Mason City Times*, February 4, 1891.

Charles Willard Hayes, date unknown. (U.S. Geological Survey photograph) [Portraits, #301]

I would respectfully request that one of the geologists of your survey be allowed to accompany my proposed expedition in that capacity, his researches to accrue to the Survey, matters of general or popular interest to be allowed me for literary purposes, the Survey to continue his salary while I am to pay all other expenses resulting from his connection with the expedition.[24]

24. Schwatka to Powell, February 20, 1891. USGS. "Register to Letters Received, 1879–1901." NARC., Wash. D.C., 20408.

The S.S. *City of Topeka* on which Lt. and Mrs. Schwatka and C. W. Hayes traveled to Alaska in 1891 is shown here anchored at Muir Glacier. (Winter and Pond Collection, Alaska State Historical Library) [PCA 87–1736]

Powell agreed and directed his Chief Geologist, Dr. G. K. Gilbert, to canvass his staff for volunteers. Charles Willard Hayes, a young geologist with the survey, jumped at the chance.

Hayes had never been on a lengthy exploration, but his superiors were impressed with his ability as a leader, and knew him to be an outstanding field geologist. For Hayes it was the chance of a lifetime—to take part in a major expedition with a famous explorer and to publish on the geology and geography of an undescribed region. Hayes was well qualified. He was born in Granville, Ohio, in 1858, and attended Oberlin College, where he graduated in 1883. After a year spent teaching high school he entered Johns Hopkins University for graduate work and received his doctorate in 1887, with chemistry as his major subject and mineralogy and geology as minor subjects. After graduation Hayes joined the U.S. Geological Survey where he assisted Dr. Israel C. Russell on a survey of the southern Appalachian Mountains.[25]

---

25. Dr. Israel C. Russell was an experienced Alaskan explorer, having made a reconnaissance of the Yukon Valley in 1889 and carried out extensive glacial studies in the Saint Elias region in 1890. He failed in an attempt to reach the top of Mt. St. Elias in 1891. It was through his association with Russell that Hayes first became interested in Alaska (Brooks 1916, p. 100).

Off to Alaska

By spring Schwatka felt recovered enough to set off for Alaska, thanks to his strong constitution and Ada's good nursing. That he would still consider such a rigorous undertaking in view of his health problems and poor physical condition says a great deal about his courage and determination.

Schwatka, Ada, and Hayes traveled to Puget Sound by train and in early May boarded the Pacific Coast Steamship Company's steamer *City of Topeka* in Tacoma, Washington Territory. They arrived in Juneau a few days later. Here they arranged for the expedition's freight to be unloaded and stored, then continued on in the *Topeka* to Glacier Bay, Sitka, and Killisnoo, cruising the popular sightseeing loop among the islands of southeast Alaska as the ship carried passengers and freight to canneries and settlements along the way. On their return to Juneau on May 9, they disembarked and began preparing for the overland journey.[26]

In planning the expedition Schwatka hoped to explore as much unmapped country as possible. His original plan had been to enter the Interior by way of Chilkoot Pass as he had done in 1883, continue in folding boats down the Yukon River to Fort Selkirk, then travel overland through the unmapped region between the Yukon River and the eastern branches of the Copper River. But on board ship he was approached by Judge L. L. Williams and other prominent Juneau passengers who urged that instead he begin the expedition in Juneau, entering the Interior by way of the Taku River.

By 1891 Juneau had become the center of gold mining on the Alaskan coast, and the most important outfitting point for prospectors heading inland. Most entered by the "miner's route," over Chilkoot Pass at the head of Lynn Canal, a hundred miles by water to the northwest. But a few reached the Yukon Basin by way of the Taku River, located only a few miles southeast of Juneau, the route long used by the Inland Tlingit. If the Taku could be developed as a major trail, Juneau's future position as entry point for the Interior would be assured.

In Juneau Schwatka was entertained by other local citizens who continued to urge him to consider the Taku route, and offered substantial support for the expedition. It was a chance to explore and survey a second unmapped region, and Schwatka agreed.

---

26. *Juneau City Mining Record,* May 14, 1891, p. 5.

Juneau waterfront scene, 1894. The S.S. *City of Topeka* is moored to the dock at left. (Winter and Pond Collection, Alaska State Historical Library) [PCA 87—78]

Of great benefit to the expedition was the addition of a third member, Mark C. Russell, a local miner and prospector who had entered the Interior by way of the Taku River the previous year.[27] Regarding Russell's trip Hayes later wrote:

> The whole of the route from Taku inlet to the Lewes was traversed in the spring and summer of 1890 by a party of eight miners, among whom Mark Russell, a member of our party, was a leading spirit. They started from Juneau before the ice was out of the river, hauling their outfit on hand-sleds so long as the snow lasted, and then packing them. It required eighty days to reach the lake, [Teslin] where the party built a number of boats. After prospecting the Nisutlin and other streams on the eastern side of Ahklen valley they went down the Teslin and back to the coast by Lewes river and Chilkoot pass. This is an example of the many unheralded expeditions which the Alaskan prospectors have carried out, facing dangers and privations which appear incredible to one who is not familiar with the men themselves. Less arduous or novel

27. The prospector Mark C. Russell was not related to Dr. Israel C. Russell of the U.S. Geological Survey.

expeditions have brought fame to explorers better versed in the art of advertising than these unassuming miners.[28]

As with so many prospectors who left no written account, little is known of Russell. He was a capable and pleasant companion, well thought of by both Schwatka and Hayes. Just why he agreed to join the expedition is unclear. If he was paid, there was no mention of it by Schwatka or Hayes. Quite likely he was willing to go in return for his keep and the chance to see and prospect new ground. Russell's knowledge of the country, his experience in dealing with the Natives, and his expertise as a boatman proved essential to the success of the expedition.

While waiting for the spring breakup of ice on the Taku River, Schwatka and Ada met with influential citizens and generally tended to public relations. Following an interview with Schwatka concerning his plans, the *Juneau City Mining Record* reported:

> The patron of the expedition is "The New York Ledger," which adds another name to the already long list of enterprising literary patrons to expeditions equipped for the purpose of geographical research and exploration. This is especially true in America where so much unexplored country has been opened and made known through the instrumentality of the literary world as patrons, as well as distributors of the useful and interesting information acquired thereby.[29]

Hayes saw to scientific matters, including arranging with Reverend Eugene S. Willard to record daily sea-level barometric observations necessary for later correction of elevations obtained from aneroid measurements.[30] Russell bought supplies and put equipment in order. At the Taku Tlingit village on Douglas Island, across Gastineau Channel, they hired Natives as guides, packers, and boatmen for the first leg of the journey and arranged for use of a canoe.[31]

---

28. Hayes 1892, p. 119. Hayes no doubt had the flamboyant Schwatka in mind in making this statement.

29. *Juneau City Mining Record*, May 21, 1891, p. 1–2.

30. Reverend Eugene S. Willard was Juneau's Presbyterian minister. Schwatka had met Willard and his wife, Caroline, in 1883, when they were presiding at Haines Mission, near Chilkat (Schwatka 1894, p. 54).

31 The Taku Tlingit, a coastal Native group, maintained close family and trade ties with the inland Tlingit whose territory included the upper Taku River, Teslin Lake, and the drainages of the Nisutlin and Big Salmon Rivers in the Canadian interior (McClellan 1975). Schwatka's spelling, T'linkit, is retained in his narrative.

View looking up Seward Street, Juneau, Alaska, in the 1890s. (LaRoche photograph, W.L.R. Collection, Alaska State Historical Library) [PCA 95–57]

The expedition left Juneau by canoe on May 25, 1891, after receiving news of the breakup of ice on the Taku. Ada left Juneau soon after on the steamship *Mexico* for Puget Sound, then continued on by train to their home in Rock Island. The *Mexico* stopped in Sitka on May 29 and Ada provided a reporter with news of the expedition's departure. The Sitka *Alaskan* on May 30, in announcing the departure reported: "The patron of the expedition is *The New York Ledger*, and the expedition will be known as, 'The New York Ledger Expedition.'"

The expedition included three white men—Schwatka, Hayes, and Russell, and seven Alaska Natives. They paddled down Gastineau Channel, then made their way up Taku Inlet and Taku River to the head of canoe navigation at the village of Nakana. From here they packed overland to Teslin Lake (Ah'k-lain), arriving on June 15. The Native packers returned to Juneau, taking with them copies of scientific records, specimens of plants collected by Hayes,[32] and a letter from Schwatka to his backers in Juneau, which was published in the *Juneau City Mining Record* ten days later.[33]

---

32. For a description of plants collected, see Cummings 1892.

33. *Juneau City Mining Record,* June 25, 1891, p. 1.

View from the hill behind Juneau, looking across Gastineau Channel to Douglas and the Treadwell Mine, 1896. The smoke is from the mine smelter. (Winter and Pond Collection, Alaska State Historical Library) [PCA 117—18]

In his letter Schwatka described the trip up the Taku, playing down the difficulties, and offering his opinion that the river might someday be feasible as a major route to the Interior. Whether the account satisfied his backers in Juneau is not known but the letter made front-page news, adding to the general feeling of optimism in town.

At Teslin Lake, Schwatka, Hayes, and Russell set up folding boats that had been packed in from the coast, and continued their journey toward the northwest, down Teslin Lake and Teslin River, to its junction with the Yukon, then down the Yukon to the site of old Fort Selkirk. Here they again hired local Natives as guides and packers and traveled overland through unmapped country to the head of White River and across Skolai Pass to the Nizina River. After building a makeshift boat they ran the Nizina's rapids down to the Chitina River and on to Taral,

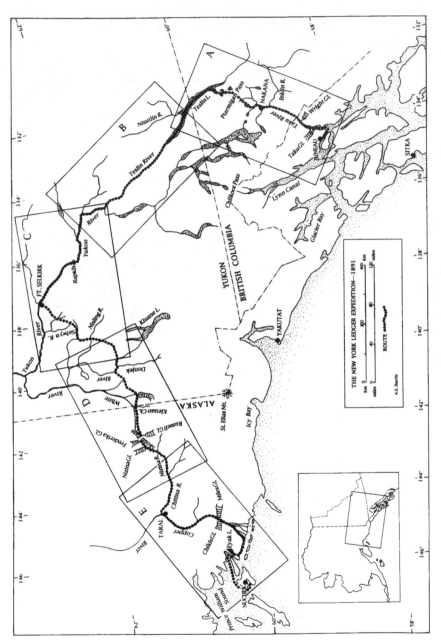

Map 1: Route of the New York Ledger Expedition of 1891 from Juneau to Nuchek, Alaska–May 25 to August 24, 1891–with key to detailed route maps. Topography based on U.S. Geological Survey, State of Alaska, Map B, 1986.

an Atna Native village on the Copper River near the present town of Chitina, an area previously explored and mapped by Lieutenant Henry T. Allen.[34] Here they joined the Ahtna chief, Nikolai, and a party of Natives, and floated in their bidarka down to Prince William Sound, arriving at Alaganik, a fishing station on the west side of the Copper River Delta. After waiting out a storm, they rowed and sailed up Eyak River in the company of two local fishermen, to the fishing station on Eyak Lake. A short hike on the fish tramway across the isthmus brought them to the Pacific Steam Whaling Company's salmon cannery at Odiak, on Prince William Sound near the present city of Cordova.[35] At the cannery they boarded the *Salmo*, a steam cannery tender, for their planned destination of Nuchek, a village of 150 residents on Hinchinbrook Island, where they arrived on August 24. Here they found mail waiting, the first received since leaving Juneau just short of three months before.

Their arrival at Nuchek came one day too late to catch the North American Commercial Company's mail steamer *Elsie* on her monthly run to Sitka, and there was nothing to do but wait. Hayes and Russell put the time to good use—cruising Prince William Sound in the *Salmo* at the invitation of Captain Humphrey, and hunting ducks on the Copper River Delta.

Curiously, Schwatka fails to mentions a word of his activities during the month-long wait. Apparently he remained at Odiak to write and to recuperate from the journey. His state of health is unknown, but it was evidently not good. The men caught the steamer *Elsie* on September 21, and arrived in Sitka four days later, in time to take the *S.S. Mexico* back to Puget Sound.[36]

The *Mexico* passed through Juneau on September 29, where the editor of the *Juneau City Mining Record* tried unsuccessfully to interview Schwatka. On October 1 the paper reported:

> ...The editor of the *Record* did not interview Mr. Schwatka as he was not in his normal condition, but it is said that he accomplished more than he anticipated. The entire distance traversed covered 1,500 miles

---

34. Lieutenant Allen explored and mapped the Copper and Chitina Rivers during his expedition of 1885 (Allen 1886).

35. Moser 1899, p. 131.

36. *Alaskan*, September 26, 1891, p. 3.

of which 700 were through an entirely unexplored country. Important geographical discoveries were made by the expedition and traces of mineral were observed everywhere...[37]

Just why Schwatka could not be interviewed was unclear, and neither Hayes' nor Schwatka's narrative offer any clues. It seems quite likely, however, that the hardships of the journey had aggravated his stomach problems and that the failed interview was owing to his condition as a result of the painful attack.[38]

The *Mexico* arrived in Tacoma on October 8, after some delay. The Tacoma *Morning Globe* of October 9, 1891 (pg. 6), described the trip:

> The steamship *Mexico*, several days overdue, arrived from Alaska yesterday afternoon after a difficult passage. She experienced alternately heavy fogs and fierce gales during the whole trip down. At Sitka she was delayed forty-eight hours, unable to pass out of the harbor on account of a gale blowing outside. She was also compelled by gales to lay to for forty-eight hours under the south shore of Prince of Wales Island. Another cause of considerable delay was that the steamer put in at nearly every cannery along the coast. The fishing season having about closed up there, she brought down the last of the pack and the laborers from several canneries. Her cargo included 21,000 cases of salmon, 300 sacks of ore, 100 tons of guano, 150 barrels of oil and $80,000 of bullion. She carried 140 passengers, all told, sixty of them being Chinamen in steerage.

Soon afterward Schwatka made a trip east to New York, stopping over in Des Moines, Iowa, to visit friends. A reporter who interviewed him noted that the expedition had taken its toll:

> The fatigues of a summer spent in exploring the unknown recesses of Alaska and the dangers incident to it have made their impression even upon the robust frame of Lieut. Schwatka. He bewails the loss of about 60 pounds of good flesh, which has been lost on the route, but

---

37. *Juneau City Mining Record*, October 1, 1891, p. 1. The editor of the *Mining Record* later complained of incomplete and unsatisfactory reports and offered his opinion that the people of Juneau had been hoaxed in that they would probably gain no benefit from the several hundred dollars invested in the expedition. *Juneau City Mining Record*, October 29, 1891.

38. One may only speculate as to whether Schwatka's well-known drinking problem may have contributed.

which he confidently hopes to regain under the pleasanter conditions of life in a more temperate zone.[39]

Why Schwatka traveled to New York is unclear, but it likely involved publication of his narrative, which was serialized in eighteen installments in *The New York Ledger*, between March 19 and August 13, 1892.[40]

Following his journey east, Schwatka returned to the Pacific northwest where he hoped to make a successful speaking tour. But by then his drinking had become a serious problem, which he finally recognized as a disease that he intended to overcome. In a lengthy newspaper article he wrote:

> There is a very large class of people who drink within the limits of never showing it, or at least never showing it in their business, and the world calls them "moderate drinkers" and "those who do not drink to excess," when the fact is, that it more or less seriously interferes with their business, and especially if it be mental labor. It is oftentimes harder for this class to stop drinking than the man who goes on a spree occasionally and is dubbed a "drunkard," while the other is not….There has been a great deal said by all sides as to whether the liquor, opium and other similar habits can be called diseases…some extremists claiming that the subject is wholly one of morality and the others that it is wholly one of pathology, or purely a case of diseased conditions. Like a good many other discussions, I think both sides are right in a greater or less degree. In the early part of acquiring any of these habits, there is no doubt more of the moral side to the subject and could the person concerned look ahead and clearly comprehend the true effect of the habit, this fact alone would be sufficient to put a quietus to it. In its later stages it is as clearly a disease as any other morbid condition of the system or any of its parts have ever shown.[41]

Schwatka resolved to stop drinking by taking the so-called "Keeley Cure," an institutional regime for alcoholism and drug addiction promoted by Dr. Leslie E. Keeley that included injections of bi-chloride of

---

39. The story was carried by *The Daily Argus*, Rock Island, Illinois November 23, 1891. There appears to be no record of Schwatka visiting his home in Rock Island during this trip.

40. Schwatka 1892b.

41. *The Daily Ledger*, Tacoma, Washington, July 25, 1892.

gold, amorphin, strychnine, and other alkaloids.[42] The cure was popular in the eastern United States but the only Keeley Institute in Washington Territory was located in Olympia, and it was there that Schwatka went for treatment. About the dangerous treatment Schwatka wrote:

> I was heavier dosed than most of those who have been reported as injured thereby. My constitutional aversion to medicine is great, having taken 180 grains of choral [chloral] to produce sleep, while forty grains have proved fatal; and also a hypodermic injection of morphine five times past the fatal point, and other things in proportion. In no case did I feel the remotest sign of unpleasant symptoms in the Keeley cure....There is a little 'heaviness in the head,' felt during the treatment, that oftentimes disappears in its course, and always afterward.[43]

By July Schwatka pronounced himself "cured" and resumed his lecture tour of Washington, appearing in Olympia, Tacoma, Port Angeles, and Seattle. He wrote articles for local newspapers, offering advice on encouraging trade with Alaska, the feasibility of a railroad link to Siberia, and anecdotes on his arctic exploration and his life on the plains. He also worked with the Commercial Club of Tacoma, submitting a plan to explore Mount Rainier and write a book on the subject as part of an effort to help lobby for establishment of a national park.[44]

Schwatka appeared to be back in form with a busy schedule. His cure was apparently successful insofar as his drinking habit was concerned, but there were signs that all was not well. A reporter later noted:

> ...according to his friends, while in Tacoma he did not drink a drop. While here [in Seattle] he gave repeated lectures on his famous arctic exploring tours, but his efforts failed to draw large crowds. On the whole it is understood that he was peculiarly embarrassed and discouraged...[45]

---

42. Preble 1932.

43. *The Daily Ledger*, July 25, 1842.

44. *The Daily Ledger*, September 14, 15, 18, 21, 1892. Newspaper accounts referred to the mountain by the native name, *Tacoma*, rather than to the name *Rainier*, which Captain George Vancouver had bestowed. A bill was introduced in Congress in 1894 to create the Washington National Park, which included Mount Rainier. Early attempts were unsuccessful, but in 1899 a revised bill was passed by Congress and signed by President McKinley establishing Mount Rainier National Park (Brockman 1959, p. 59).

45. *The Seattle Post-Intelligencer,* November 3, 1892.

Schwatka left Seattle on September 25, for Portland, Oregon, where he had relatives.[46] While in Portland he continued lecturing, boarding with an old friend, Dr. T. L. Nicklin, at the G. W. Bolter residence on Third Street. During this time his stomach problems failed to improve. He was in frequent pain, alleviated by the continued use of laudanum. On the evening of November 1 the two friends dined together. Schwatka was in his usual jovial mood but complained of especially severe pain in his stomach and they parted company at 7 P.M. Later that evening the pains grew worse and Schwatka purchased two ounces of laudanum at a local drugstore. He was scheduled for a lecture at 9 P.M., but to his relief found the lecture had been canceled.

The next morning a policeman found him slumped in a doorway on First Street near Morrison, a half-filled bottle of laudanum by his side. The officer thought Schwatka was drunk, and he was taken to the city jail, but after an examination he was sent to Good Samaritan Hospital, where he died without regaining consciousness—from what was described as an accidental overdose of laudanum.[47]

News of Schwatka's death was telegraphed to Ada, in Rock Island, and arrangements were made through his sister, Mrs. T. H. Reynolds, for his burial in the family plot in Salem. Ada did not attend the funeral.[48]

Hayes returned to his home in Washington, D.C. where he resumed his duties with the Geological Survey. He presented a paper on the expedition before the National Geographic Society on February 5, 1892, and his account was published in the May 15th issue of *The National Geographic Magazine.* His account dealt mainly with scientific aspects of the expedition, in particular with the geology and geography of the regions traversed, and included detailed maps based on his track surveys. In deference to Schwatka's intention to publish his personal narrative, Hayes offered only a brief narrative of the expedition. His

---

46. Family members residing on the west coast included two sisters in Salem, Mrs. Amelia Strong and Mrs. J. D. Jordan; three sisters in Portland, Mrs. Thomas H. Reynolds, Mrs. M. C. Cross, and Mrs. Annie Hunsacker; a brother, Augustus, in San Francisco and an uncle, Ned Schwatka, in Eureka, California (*The Morning Oregonian,* November 3, 1892, p. 10).

47. An autopsy revealed no organic lesions sufficient to account for his death, and his death was attributed to an overdose of laudanum (*The Morning Oregonian,* November 3, 1891, p. 10). Following an inquest, a coroner's jury found his death to have been accidental (*The Morning Oregonian,* November 4, 1891, p. 10). News of his death was carried by newspapers throughout the United States. (*See also* Sherwood, 1979.)

48. *The Morning Oregonian,* November 4, 1982, p. 10.

published account was of great scientific value, and has provided the only readily available source of information regarding the expedition.

Hayes continued his work with the Geological Survey, and the following year married Rosa E. Paige of Washington. He continued work on the geology of the southern Appalachians, and in 1902 turned his attention to the geology of petroleum, making personal investigations in Louisiana and Texas and directing geologists in the California and midcontinent oil fields. By 1905 he had taken on the administrative duties of Chief Geologist of the U.S. Geological Survey, and two years later the post was made official. In 1910 Hayes was appointed as geologist to the Nicaragua Canal Commission, later taking on similar duties in the Panama Canal Zone.

Hayes had earlier taken furlough from his government post to do consulting work in Mexico. His knowledge of stratigraphy and his ability to locate oil deposits led to an offer to become vice president and general manager of the Aguila Oil Company. He resigned from the Geological Survey in 1911 and with his family moved to Tampico. Two years later revolution forced them to leave the country. They returned to Washington, D.C. where Hayes died of cancer in 1916. He was survived by his wife and six children.[49]

What became of Mark Russell is not known. Like so many other prospectors who failed to strike it rich, his name does not appear in connection with stories of sudden wealth. Some evidence suggests that he may have remained in Alaska, although official records have not been found.[50]

Schwatka's Narrative Lost

The New York Ledger Expedition was Frederick Schwatka's most ambitious Alaskan expedition, all the more remarkable because of his physical problems at the time.[51] Yet his personal narrative has remained lost for over a century. How could this have happened?

---

49. Brooks 1916.

50. Mark Russell, a member of the 1887 Pioneers of Alaska, is listed as a pall-bearer at the funeral in Juneau of a fellow member, China Joe (*The Daily Alaska Dispatch,* May 22, 1917).

51. The noted Alaska historian, Morgan Sherwood, wrote of Schwatka's New York Ledger Expedition: "In terms of original discoveries, natural obstacles, and contributions to science this 1891 expedition was the most notable journey Frederick Schwatka made in Alaska..." (Sherwood 1992 p. 143).

By 1891, at the start of his third Alaskan expedition, one sponsor had signed on—Robert Bonner, owner and publisher of *The New York Ledger.* Bonner had built the Ledger's reputation by paying well for work by prominent writers, and his generosity had attracted such literary names as Dickens and Tennyson, along with others now forgotten.[52]

Whether Schwatka tried to enlist additional support from newspapers is not known, but Hayes appears to have thought that support was more widespread. In his article published in *The National Geographic Magazine* in May of 1892, he referred to a "syndicate of newspapers" sponsoring the expedition, evidently believing that such backing was assured.[53] His reference to a "syndicate" implied that prominent coverage by the press would be forthcoming, but such coverage failed to materialize. This lack of press coverage, coupled with Schwatka's well-publicized death soon after, led historians to assume that Schwatka died before he could publish his narrative. The brief announcements in the Alaskan newspapers identifying *The New York Ledger* as sponsor were overlooked.

Schwatka's narrative was styled to appeal to a wide readership, but its one-time serial publication in *The New York Ledger* limited its audience. Sandwiched as it was between such contemporary romances as *Morris Julian's Wife, Cynthia Wakeham's Money,* and *The Brigand's Hand,* it also failed to reach the contemporary audience of scientists and geographers he would have wished. Robert Bonner died in 1899 and the *Ledger* ran into financial problems. After several mergers it ceased publication in 1903 and Schwatka's narrative, along with the *Ledger* itself, were soon forgotten.[54]

Because of Hayes' publication in the prestigious *The National Geographic Magazine* and his later standing as a prominent scientist and geologist, most of the credit for the expedition has been his. This credit

52. Edwards 1899; Lee 1957.

53. Hayes 1892, p. 118.

54. Even Eliza Ruhamah Scidmore, one of the most renowned chroniclers of Alaskan affairs at the time, was unaware of Schwatka's narrative. In an 1894 review of recent explorations in Alaska, she wrote: "His untimely end prevented his publishing the narrative of a journey as hazardous and important as any he ever attempted." (Scidmore 1894, p. 176) Subsequent writers have similarly erred in believing that Schwatka died before he could publish (Sherwood 1992, p. 143; Webb 1985, p. 118), or were apparently unaware of Schwatka's published narrative (Brooks 1973, pp. 281–282; Wright 1976, p. 222).

Title page of *The New York Ledger* dated March 19, 1892.
This issue featured the first installment of Frederick
Schwatka's narrative. (Author's collection)

is well deserved. Hayes contributed substantially to geography by first
mapping the Taku route from tidewater to Teslin Lake and the route
from Fort Selkirk to the Chitina River. He named many physical
features, usually recording Native names, but sometimes substituting

his own place names.[55] His many geological observations included determining and first reporting the location of the copper deposit at the head of the White River long used by the Native people.[56]

However, it was Schwatka's vision and leadership that made the expedition possible, and his many observations that continue to make his narrative interesting and important. Schwatka wrote first to entertain his readers, but also to bolster and maintain his reputation as a scientist, geographer, and explorer. His vivid and often rambling prose tended to alienate serious scholars who dismissed much of his writing as merely efforts to titillate his readers. But, in Schwatka's defense, it must be noted that he wrote with humor for the general public. He knew what his readers wanted and he wrote to sell. Because of his popularity, his writing and lecturing did much to acquaint the general public with Alaska and the Yukon Basin.

Penned by a scientist trained in geology, Hayes' published writing conveyed the essence of his geological and geographical findings clearly and concisely, and his heretofore unpublished journal gives a detailed account of the expedition.[57] Taken together, Schwatka's and Hayes' accounts offer a fairly complete look at a remarkable expedition, a glimpse of the day-to-day life of the Native people they met along the way, and some insight into the personalities of the expedition members.

Schwatka named or renamed numerous geographical features and has been criticized by geographers and some modern writers for ignoring Native place names or names already in common usage by prospectors and miners. At least some of his renaming was an effort to reduce confusion caused by the difficulty white men had in distinguishing Native names.[58] That so many of his place names remain on maps

---

55. Among other features Hayes named Wellesly Lake in Yukon (known to the natives as Cho-ko-mon), for the college in Massachusetts where his sister, Ellen, was professor of mathematics, and Wright Glacier, in Alaska, for Professor G. F. Wright, his major professor at Oberlin College.

56. Hayes 1892, pp. 143–144.

57. Hayes' journal, consisting of two hand-written notebooks, is on file at the office of the U.S. Geological Survey, Branch of Alaskan Geology, Anchorage, Alaska 99501.

58. For example, Schwatka (1894, p. 189) noted that during his expedition of 1883 his Chilkat companions used the name Tahk-heen'a for one river and Tah-heen'-a, for a second river. To avoid confusion, Schwatka retained the native name Tahk-heen'a (now Takini) which Dr. Aurel Krause had noted in 1882 (Coutts 1980, p. 259) and renamed the Tah-heen'-a, the D'Abbadie, after the French explorer. The D'Abbadie is now known as the Big Salmon River (Coutts 1980, p. 259).

today is due to his penchant for naming physical features to honor prominent geographers and scientists—names that later geographers were
reluctant to change. For example, George M. Dawson, Chief of the
Canadian Geological Service wrote:

> ...he has completely ignored the names of many places already well
> known to miners substituting others of his own invention....Strict jus-
> tice might demand the exclusion of all these on the definitive maps now
> published, but...especially in view of the scientific eminence of some of
> the names which he has selected, it has been decided to retain as many
> as possible of these.[59]

While some criticism of Schwatka is probably deserved on this score,
his feats of organization and leadership are remarkable and his writing
and lecturing entertained and informed a generation of Americans.
Whether through good luck or design, his ability to recruit able associates added a great deal to the scientific value of his expeditions.

Judged by standards of today Schwatka may be criticized differently—on grounds of racial prejudice and insensitivity. Although a keen
observer, his attitude toward Native Americans was patronizing and his
views were colored by military life on the frontier of western America.
What was once considered "humor" we now recognize as being based
on prejudice that at the time was widely shared. This offers a jarring
note today and helps illustrate the great cultural gulf between
Euroamericans and Native Americans a century ago. The extent of our
discomfort today reflects how far cultural attitudes have changed.

It is obvious in Schwatka's writing that he depended on Native
Americans for their knowledge of the country, their help in packing,
and their willingness to supply food and help when needed. His respect
for their business acumen is apparent, as was his respect for their physical stamina and compatibility with their environment. Without their
help his expeditions would not have succeeded.

Frederick Schwatka is buried in Salem, Oregon, in the family plot at
the old Rural Cemetery on Commercial Street. His small marble head-
stone describes how he thought of himself and how he wished to be
remembered. It reads, "Frederick Schwatka—Explorer."

---

59. Dawson 1987, p. 143B.

NARRATIVE OF FREDERICK SCHWATKA

1892

FIRST LETTER

On May 7th, 1891, the Pacific Coast Steamship Company's vessel *City of Topeka* crossed Dixon Entrance and entered Alaskan waters, leaving behind those of British Columbia. All of these waters, for a stretch of over a thousand miles along the Pacific coast, are but deep inlets and channels—huge salt-water rivers, so to speak, that, cutting in every direction, make a vast network of islands, picturesque in the extreme, and not even yet wholly explored.

One of the most freely navigable of these many channels, or rather a series of channels connecting with each other and running parallel with the coast, has been selected by a tourist line of steamers, and every summer its vessels are crowded with pleasure-seekers attracted by the magnificent Alpine scenery, the restfulness of an ocean voyage without its discomfort of sea sickness, the unusual phenomena of Arctic glaciers and icebergs, and the most delightful and invigorating of climates.

This is fairly well known as the "inland passage to Alaska," and is yearly becoming more and more popular with the summer tourists seeking rest and recreation. In fact, one of the peculiar features of this southwestern section of Alaska, and one that can only be appreciated by the explorer or the hardy prospector or trapper, is the picnic-like pleasure with which they can travel the coast and its estuaries, only to plunge into the roughest of "roughing it" as soon as the salt-water channels are left and the Alpine interior is essayed.

On board the *City of Topeka* was an expedition, the patron of which was the New York *Ledger,* that had this prospect ahead of it for at least the summer of 1891, and with no small chances of wintering in the bleak interior should any of the not unusual exploring casualties determine such an undesired event. As one of our main objects, however, was to cover as much unexplored country as possible, our plans were not of that rigid nature imposed upon a traveller who has to make a fixed point, whatever may be the intervening obstacles, or consider his project more or less of a failure. Nevertheless, in a general way, we had promised our patrons a crude plan, which may be briefly summed up as follows: To disembark from the Alaskan steamer at Chilkat the north-ernmost point of the inland passage, make our way over the Chilkoot trail through Perrier Pass to one of the heads of the Yukon River and down that stream, in folding canvas boats, to some point between Fort Selkirk and the mouth of the White River where Indians could be best obtained for packers to make a three-hundred to four-hundred-mile portage to the Atna (Copper) River or its eastern branches, from which civilization could be easily reached down that stream and out of Prince William Sound. We had occasion to change this plan, so as to cover more unexplored country than intended, and it came about in this way: On board the steamer were some residents of Juneau the mining, mer-cantile and commercial metropolis of Alaska—"the largest city in the largest territory of the largest free republic in the world," as they used to patriotically put it. They were naturally interested in all explorations of the country, especially those backed by the press as patrons, in that a wider diffusion of the results were generally given, and oftentimes in a shorter time, than by the limited circulation of government reports. They were especially concerned in the way by which I intended to reach the head of the Yukon River, as nearly all the placer gold mines of the upper portion of that great inland stream were more or less tributary to Juneau as an outfitting point.

In breaking through the Alaskan coast range of mountains by the northernmost trail—the Chilkoot—I was taking a pass first explored by the Krause Brothers, sent out by the Bremen Geographical Society of Germany, and over which I had gone a year later in command of the United States Alaskan Exploring Expedition of 1883; so it was not only fairly well known, but it would be also old matter for my own pen. Still, it was deemed by every one as the only practical pass to reach the headwaters of Alaska's greatest river, despite its obstacles over glacier-clad paths that reached to some 4,000 feet above the level of the sea.

Near Juneau, a deep inlet, in characteristics similar to the many chan-nels of that region, stretched well into the interior. It is called the Takou River, and at its head comes in a great glacier, and near by a swift river of considerable size. From near the headwaters of this stream it was known that one of the principal branches of the Yukon could be reached, and it was thought to be by a trail much better, even if longer, than the one by the old Chilkoot paths, while the new Yukon branch was known to be less obstructed by rapids and other impediments than by the branch which led out from the Perrier Pass. In fact, it was surmised that a powerful light-draught river steamer could ascend to the head of canoe navigation as marked by the Takou Indian traders, while a pack-mule trail could be made to the Yukon tributary, and this again, on its part, would give uninterrupted summer navigation of that great stream clear through to Bering Sea. If all this obtained, it was clearly a matter of sufficient mining, fur-trading, or other commerce on the Yukon, either for the present or future, to justify opening such a line of communication, while it would be no small feather in the cap of the New York Ledger Expedition to be able to do this preliminary pioneering.

As far back as their legends extend the Takou Indian fur traders of the interior had passed backward and forward over this trail, packing on their backs and canoeing on the rivers. A party of adventurous prospectors had ascended the frozen river with sledges and made the Yukon beyond, but they had been about three months making it and had reported it correspondingly difficult and unavailable—a deduction to which they gave practical illustration by re-entering the valley next year by the old Chilkoot trail. Some of these miners had said, however, that they had had no Indian guides, following only an old native sledge track, and been forced to abandon even this on account of its erratic and uncertain course, striking boldly out on their "own hook" for the mythical river. The many other reasons for or against this trail can be made clearer as we travel over it in a later letter. Suffice to say we discussed it, and Judge L. L. Williams, ex-commissioner of the Juneau district under President Cleveland, took hold of the matter enthusiastically and yet practically, so that when we arrived in this lively little metropolis of our distant colony a corresponding interest was soon worked up, the citizens pledging everything they could to make the enterprise a success, the most important being the moral backing which goes with the united action of any assemblage, as will appear somewhat further on. And so we come to disembark at Juneau for almost immediate explorations, instead of Chilkat, where we were still distant from original investigation, following the route we had proposed.

But we did not disembark at once, for this Alaskan port, on account of its commercial importance, is touched again by the steamer as she returns, and we made this short "round trip," of a few days, as a sort of preliminary pleasure to the unavoidable discomforts of the later exploration that would last as many months.

And this, too, suggests that a very brief description of the "Alaskan inland passage," as we found it, would not be inappropriate here. The steamer leaves the Puget Sound country and Pacific British ports, connecting with all transcontinental railways, the Canadian Pacific being the most direct, and traverses British Columbia inland waters until Alaskan channels are reached.

The first point we touched was Tongass, where a salmon cannery, burned last year, used to be located.

This is a favorite hunting locality for the Columbia black-tailed deer, the only species found in Alaska, and these confined to the many islands of the southwest, the wolves, it is said, running them off the mainland. One of the officers of the ship told me that in 1890 over forty deer carcasses were taken here at one trip, a hunter having killed seventy of them in ten days. These forty carcasses were taken to Juneau, and brought but two dollars each in the markets that were thus overstocked. After Tongass we are touching at salmon canneries for a day or two, amusing ourselves at watching the "puffing pigs," the porpoises and the millions of sandpipers that at this season line the shores of this great inland salt-water network of channels. Wrangell, the supply point for the Cassiar placer mines of the interior, is reached the 8th and Juneau the 9th of May, where our freight disembarked, and we continue on. All of the high mountains are yet covered with snow, but the weather is very fine, and all enjoy the magnificent character of the almost Arctic scenery. Leaving Juneau, we look into the deep, dark Takou inlet with its gleaming-white icebergs floating on the green water, which is flanked by rock-ribbed mountains surmounted by snow and glacial ice.

The party wakes up in Glacier Bay next day, the *Topeka* being the first vessel of the season to enter its ice-laden waters. There is no commerce in this bay, and it is only entered in the spring and summer months, to accommodate the many tourists then flocking to Alaska. The Alaskan traveller coming by the Canadian Pacific Railway, with its great glaciers almost reaching the very track, and its snow-clad Selkirks, coupled with a day or two in Glacier Bay of Alaska, with its icebergs and mighty mountain scenery of the far away Fairweather group, sees really more of the Arctic regions, or its equivalent, than many travellers who actually pass beyond the Arctic circle itself for only the summer.

Juneau, Alaska, 1891. (Photograph by F. Jay Haynes, Haynes Foundation Collection, Montana Historical Society, Helena, Montana)

And surely, our encounter this morning with the ice was Arctic enough! The brash ice and bergs swung into dense "packs"—as the polar ice-master would say—that, extending from shore to shore, made it seemingly impossible to reach our coveted destination, the front or ice-front of the great Muir glacier, the parent source of all these floating obstacles and the most imposing sight on the tourist route to Alaska. But Captain Wallace put the *Topeka* into the "pack" as if the vessel was specially prepared for true polar exploration, and for the first time in many years I heard the old sound of ice crunching and grinding against a ship's side. As the ice closed behind us, our course was plainly visible by the red paint we left on the ice cakes; a course that we could even make out by the same means when we returned some three hours later. But at last we were successful, and the *Topeka* anchored about a mile from the glacier. I was very much surprised to see so few icebergs dropping from the glacial front, as I had so often seen in my visits before, but the mate maintained that this was nearly always the case at ebb-tide, the flood being prolific of falling ice and the birth of many icebergs.

Sitka was reached next day amidst beautiful weather that continued during our stay, which was wound up with an American wedding in the old Russian building.

There were many passengers for Juneau to attend the spring term of the Alaska court—Juneau and Sitka alternating with the spring and fall terms—and, as all the territorial officials are at the latter point, we were well crowded on the return. There were ten prisoners, and it was reported that there were six murderers among them, mostly confined to half-breeds and Indians. It gave us something to talk about, and the resulting conversation disclosed that there had never been a lynching nor a legal execution in Alaska. Until within the last few years all criminals had been sent to Oregon for trial, and a few had been executed there, but since Alaska has been given a "district" government, no heinous murder has been committed. Still, I understood from another source, probably not quite so authentic, that "looking backward," he could recall when the citizens of a certain Alaskan town played with one end of a rope and an Indian murderer of a white man fooled with the other, to the satisfaction of one side and the dissatisfaction of the other. In the very early days of Alaska, a gambler who shot a miner at Wrangell was tried by the Vigilance Committee and executed; but anyone familiar with the frontier knows there is a wide difference between such action and lynching.

Killisnoo, a picturesquely placed port, is very agreeable to the eye and abominable to the nose, for here there is found a very flourishing herring fishery and an attendant guano factory. The United States marshall had some business here with an Indian for trying to drown a native woman accused of witchcraft. I have never yet seen anything bewitching about an Alaskan Indian woman, but I suppose tastes vary. The would-be murderer had some inkling of the proposed proceedings, and had secreted himself in the Alaskan underbrush, where the proverbial needle in a haystack would be as conspicuous as a circus poster by comparison. There were not eleven prisoners from Killisnoo to Juneau. In many ways the Indians of southeastern Alaska try even yet to enforce their old rites and superstitions, and avoid the American authorities in so doing.

At Juneau we moved into a little house on the high ridge overlooking the town, and giving a most picturesque view in nearly every direction. Preparations were begun at once, but the first obstacle we ran into was the report that the Takou River had not yet broken up and, of course, we could not canoe up it until it was free from ice. It had broken up the year before on May 27th, just two weeks from the present date, so the prospects were not very inviting. This report had come from

The view down Main Street, Juneau, Alaska, 1891, and south across Gastineau Channel to Douglas Island and the Treadwell Mine. The smoke is from the mine smelter. (Photograph by F. Jay Haynes, Haynes Foundation Collection, Montana Historical Society, Helena, Montana)

some white fishermen at the Takou's mouth, while the Indians were equally emphatic that the river had broken up a week or two before, as learned from other natives of the same tribe living on its banks. As usual in such controversies, the Indians were nearer correct.

Near Juneau are two bands of Indians (both of the great T'linkit tribe, however) the Auks, to the westward, and the Takous, to the eastward; and it was from these two bands I expected to get my canoemen to the head of navigation on the Takou River, and to act as porters or carriers from there across the portage to navigable waters on the great Yukon.

The next two or three days were occupied in haggling with the Indians, whose demands, as usual, were exorbitant in the beginning. However, we soon saw that we could get a good canoe and transportation up the river at a fairly reasonable rate, but the other part of the journey was the harder to arrange for. On the 19th, a prominent Takou, a sort of sub-chief, Yash-noosh—locally known as "Johnson"—came to the rescue, and from that time forward negotiations were conducted with a little more chance of success. The benefit of the moral support of the town was now manifested, and I doubt if without it we would have been able to procure enough porters for the land trail. The Alaskan coast natives, when acting as packers for themselves or others across the passes of the coast range of mountains, usually carry about one hundred pounds on their backs, the Chilkat and Chilkoots, the best of all the T'linkit tribes for this purpose, averaging probably a little more and the others a trifle less. So hard did I find it to obtain Indians that I was compelled to compromise and take half the number wanted and "double" the loads over the trail; that is, each packer passed twice over the trail loaded and twice unloaded, of course receiving double pay, and, worst of all to me, consuming double time.

Yash-noosh displayed an energy uncommon for an Indian, which may be accounted for partially by the fact that he was a paid policeman of the town for the Indian quarters, and he knew the town was in earnest about helping the *Ledger* expedition. He even made long trips after men; and one of these extended to the mouth of the Takou River showing that the stream was clear of ice despite all reports to the contrary.

By the 24th of May there had been secured six packers, after seemingly enough parleying and dickering to have secured six hundred, and the next day was appointed for the start.

On the recommendation of some citizens who believe in the horror of the average Alaskan Indian for all legal and documentary papers, I had a huge contract drawn up, resplendent with many colored seals and ribbons, and this the natives signed by touch of pen and witnesses, while the United States District Court interpreter read aloud its contents with a solemnity equal to a death-warrant. I still retain that ponderous page of legal ludicrousness, and while admitting it may have done me much good, yet I can only compare it with the verdict of the frontier coroner's jury of which I heard, wherein, sitting on a man's body dragged from the river's bed and riddled with bullets, they concluded that the deceased had come to his death by drowning caused by water pouring in through the bullet holes.

"Robert," a brother of "Johnson," furnished the canoe—a huge two-ton or three-ton affair—that easily carried the party of ten persons and their effects of nearly a ton in weight.

The New York Ledger Expedition starting up the Taku River.
Frederick Schwatka is seated at left, Mark C. Russell is standing
fourth from left. May 26, 1891. (U.S. Geological Survey photograph by C. W. Hayes) [Hayes #335]

About noon, with the American flag hanging from the peak, we got
away in the beginning of a wind and drizzling rain-storm that, later,
made our first day's trip one of the most unpleasant of the whole expedition. An enthusiastic crowd of citizens lined the shore near the steamer's
dock, and, as we paddled away down the channel, gave us many a
hearty cheer and many warm wishes for our success.

Let us now take a hasty look at the little expedition as it started. The
commanding officer was the heaviest one of the party, and made excellent ballast in the rear part of the canoe. His avoirdupois surplus was
not so beneficial, however, in some other parts of the trip.

The scientist of the expedition was Doctor C. Willard Hayes, repre-
senting, also, the United States Geological Survey through the courtesy
of Major J. W. Powell, the head of that government bureau. To Doctor
Hayes also fell the photographic work in the main, as well as the
topographic and map-making. I had hoped to get a professional

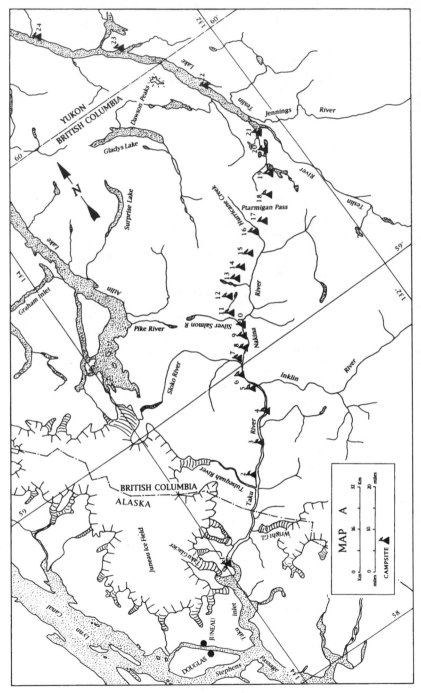

Map A: Route of the New York Ledger Expedition of 1891 from Juneau, Alaska, (May 25) to Camp 24 (June 20). Topography based on U.S Geological Survey Alaska Topographic Series: Juneau, Skagway, Taku River, and Canadian Geological Survey Map Sheets: Atlin (104N), Jennings River (104O), Teslin (105C). Glacial positions along Taku River based on surveys of 1888–1893 (USC&GS Annual Report for 1895, Map No. 22).

photographer, and felt sure of success when Mr. Landerkin, one from Juneau, made an enthusiastic application for the place. He was on the grand jury, having already been sworn in, but we both thought that a mere bagatelle as any sort of an obstacle. But the judge and district attourney thought otherwise. Both were willing enough, but there was nothing in the law (Oregon code prevails here, by act of Congress, "so far as it is practicable,") that would allow it. The only excuse whatsoever was one of severe sickness, attested under oath by a physician. "But suppose a grand juror dies?" I asked the judge. "I could not excuse him from duty on that ground," replied the judge, shaking his head. So Mr. Landerkin remained, and Doctor Hayes did double duty. There were two cameras, and I managed the smaller (4 x 5), but as it did only auxiliary duty as a "snap-shot" here and there, and was abandoned before the expedition was half over, its work was comparatively nominal. The larger, a fine "Anthony" (5 x 8), made for my Mexico work, did essentially all the work.

Mark C. Russell was the only other white man in the party, which I organized on the basis of all my previous expeditions, or just as few white men as possible, with natives to do about all the work of transportation, guiding, hunting, etc., with which they are so familiar in their own countries. Suffice to say there were six Indians as packers and "Robert" in charge of the canoe. Of them, in detail, I shall speak further on.

The canoe ride down the Gastineau channel and up the Takou Inlet would have been extremely interesting under any other circumstances than the one we had to endure—a furious storm of beating rain.

We were abreast of the great Treadwell mine in half an hour, the largest in Alaska and one of the richest in the world.

The Indians were very hilarious, evidently stimulated by the enthusiastic departure given us, and they showed their appreciation by "spurting" ferociously every little while instead of settling down to steady work.

We stopped about two in a pretty little sheltered cove for lunch, and really enjoyed ourselves for a half-hour. Robert impressed me with the fact that the tide was then out and asked me to note that when it was in not a landing place was to be found anywhere along the shore for probably stretches of many miles. The high-water mark is not always conspicuous in every nook and cove, and Robert told me of ludicrous cases where inexperienced travellers, probably wearied by long travel, had seemingly good camps at low water only to be rooted out later by the incoming tide. But along most of the shore the dense timber comes to the water's edge and high-tide mark is as sharp-cut and conspicuous as the lines of a well-trimmed hedge-fence. Every now and then this ceaseless timber tract would be broken by bright-green prairie tracts of a few acres, and on one of these bordering the shore was picturesquely perched a Takou town of ten houses, Ahk-Kwan[1] by name. It is not often that a graveyard is the most cheerful part of a place, but it was seemingly so here. The only enlivening thing in a view of somber green, rendered doubly dismal by the drenching rain, was a bright scarlet flag that fluttered from a high staff at the corner of a grave. I was told by one of the Indians it indicated that that particular native had recently died.

Every few hundred yards beautiful falls and cascades were conspicuously prominent, with their silvery contrast, on the somber mountain-sides.

About four o'clock we swung around into the Takou inlet, and our change of course now allowed us to set sail, a most happy change from paddling while in wet clothing. Along we bowled, wing and wing,[2] the ice-chunks we had met at the mouth of the inlet growing larger as we ascended it toward the great Takou glacier, until they could well be dignified with the name of icebergs; certainly so in comparison with those from the Muir glacier that are thus named. The Takou glacier ice is the bluest I have ever seen, either in or out of the arctic or Alpine regions. It was really a deep bluish-black in many places, though clear as crystal. It was quite dark when we got to the mouth of the Takou River at a half-built salmon fishery,[3] the white men sleeping in it and the Indians in the smoke-house, all of us as stiff as pokers.

---

1. The village, located near Point Bishop, has long been abandoned and the site reclaimed by the forest. Only a few vandalized graves remain. Some recent maps show it as, "Kuteha Indian Burial Grounds."

2. "Wing and wing" refers to a setting of the sail rig commonly used on Native canoes of the Northwest Coast in which a square sail was extended on each side of the canoe by means of booms or sprits. The rig is well adapted for sailing with the wind. (For a description see Emmons 1991, p. 93).

3. According to Hayes, this was the fishing station of Frank Murray and Company, located on the west side of the river about one mile below the present Hole-in-the-wall Glacier.

SECOND LETTER

When the morning of the 26th of May broke somewhat clearer, and I had a good view of the mouth of the Takou River and up its valley, I was wonderfully surprised at the extent of the stream that in some way or other I had been led to conceive was an Alpine river of comparatively diminutive proportions. It was here fully a mile in width, and seemed to hold this as far as the eye could reach up the stream. At Juneau the United States Coast and Geodetic Survey— their steamer *Patterson* being then in port—had, through the kindness of Captain Mansfield and Superintendent Mendenhall, furnished us with a map of Takou inlet and the river's mouth was thereon displayed of great width. The well-known exactness of their work in such surveys, of course, put the matter beyond cavil; but so many rivers have such enormous funnel-shaped mouths that this is no positive, or even probable, criterion as to their dimensions even a few miles up the stream. Before a week had passed over my bald head I was forced to acknowledge that the Takou was one of the great streams of Alaska and the British Columbian possessions; but more as to that later.

Just beyond the mouth of the Takou, probably two to four miles, is the great Takou glacier—really at the head of the inlet—and one of the most beautiful in all Alaska. A blind glacier seems to come in very near it on the other (or western) side from the river, while all the high hill-tops around are covered with snow and *névé*, from which more or less perfect glaciers flow down every gorge and gully.[4] Truly it was an

---

4. Since Schwatka's visit in 1891 Taku Glacier has advanced over five miles, filling the head of Taku Inlet and nearly abutting the east bank of the river. This advance, along with the deposition of sediment, has effectively moved the mouth of the river several miles downstream since Schwatka's visit. The "blind glacier" mentioned by Schwatka is now known as Norris Glacier, and continues to retreat.

Native canoe men preparing to ascend the Taku River under full sail (wing-and-wing), May 26, 1891. (U.S. Geological Survey photograph by C. W. Hayes) [Hayes #315]

Arctic sight indeed, and one calculated to frighten a novice, who probably would assume that the interior was yet worse, and correspondingly harder to travel when the river no longer afforded transportation. This deception is more or less pronounced along the whole Pacific shoreline of Alaska, where, the mountainous coast once broken through, the interior gives better travelling in many ways.

We got away about seven in the morning, a good wind behind us; but the weather unfortunately closed in again in a threatening way, spoiling our chances for photographing. As we left behind the inlet, we were now traversing a country that had never been travelled before, by pen, pencil, or photographer to give the results to the world, and we accordingly felt disappointed in the dismal weather.

There was no ice whatever in the river, showing that that danger had long since passed, and further, that no glaciers emptied into the stream along its course, as we had been led to believe might be the case by a few persons with second-hand information. This last fact was important to the Juneau people for if glaciers reached the stream and it was unnavigable to powerful light-draught river steamers, then farewell to this route for the interior, except by a costly system of bridges over a wide-water course; but no such obstacles appeared.

Sailing a few miles up the river, some four or five, a picturesque gorge or break in the mountains to the north gave us a pretty vista of the Takou glacier, lying like a white wedge between the white sides of the precipitous gorge.[5] The Indians call this pass the *Koo-dah-sāke*, or "The Fair Wind Gap," as through it such a breeze could always be depended upon for a morning's sail up the stream; the very wind we were now using.

But it was not many stretches of the stream before it became so winding, and the breeze so light, that the *Koo-dah-sāke* was not always on our side, and occasionally the Indians got out and "tracked" or "cordelled" up the bank. It was a wild, picturesque sight indeed, but it was also very slow.

About nine o'clock we passed a single Takou house on the south bank that looked like a pig-pen struck by an avalanche. Dogs kept coming out from various apertures in its sides, until no less than eight stood on the high bank and favored us with the usual canine chorus given strangers. After awhile an Indian, that looked as if he had been squeezed between two colliding avalanches, crawled out, and standing in the center of the semi-circle of dogs, put his hands to his mouth and wailed, "Hoochinoo!" This *hoochinoo* is the native Indian liquor, surreptitiously distilled from sugar and molasses, using a coal-oil can for a retort and a long kelp-stalk for a worm. It tastes of all these ingredients, plus a fair share of the flavors usually given the freedom of an Indian cabin. A prospector assured me that it was not unlike proof prussic acid, flavored with "Rough on Rats." Whether the Indian had any to sell or wanted to buy—the more likely of the two—we never knew, for we were just then enjoying a good sailing wind, which we did not propose to lose for all the *hoochinoo* existing, and we never afterward heard his voice again above the din of the dogs. I have lived a third of a century among Indians, and have seen wildernesses and deserts of dogs, but have never seen one of that breed of any use before, and this was entirely unintentional.

---

5. With the thickening and advance of Taku Glacier since Schwatka's visit, a side stream of ice from Taku Glacier, known as Hole-in-the-Wall Glacier, has filled the gorge, spilling out onto the river flat and partially blocking Taku River.

Cordelling up Taku River in a Native canoe, May, 1891. (U.S. Geological Survey photograph by C. W. Hayes) [Hayes 318]

The mountain-sides were now fairly seamed with silvery cascades dashing down their black, swarthy sides; and one bluff in particular, on the south bank—just before we reached the "Hoochinoo House"—had so many foaming down its front that I called it Cascade Bluff; though travellers at other seasons may have reason to doubt the propriety of doing so.

Near here, also, we heard our first avalanche, on this same side; but it was so far up among the lowering clouds that we might well have taken it for thunder, were it not for the facts that this noisy demonstration is almost unknown on the Alaskan Pacific coast, and the lower course of the avalanche became visible before it ceased altogether. We were evidently striking the "breaking up" season for all the watery elements, avalanches and cascades pouring down the mountain-sides, and the swift, muddy river carrying a full cargo of driftwood.

There are several glaciers along the course of the Takou, but only three or four reach the valley, and none of these, as I have already said, reach the river itself so as to discharge ice-chunks or small icebergs in its waters or to impede its valley to the ordinary forms of transportation on that side. Even a railroad could be constructed the whole length of the river on either side, I believe; but this brings us to a very important discussion much further on.

We were abreast of the first glacier, on the north side, by 9:30. It is really a double glacier, their feet joining in the valley, but in nearly all other aspects they are very dissimilar; their color, inclination, amount and character of lateral moraines, etc., etc.[6] Shortly after we passed the glacial creek evidently draining the valley occupied by these two monsters, for it was carrying the usual chalk-like silt of such streams.

Near here we also saw several hair-seals, evidently after salmon, their usual diet. They make great havoc among these fine fish, and if they could be exterminated, it would tenfold more than compensate for the mere dollar or two their skins bring in the markets. A number of nations, notably the Danes, offer a reward for their scalps in order to protect their fisheries.

Ten o'clock saw us at Klame-quah, an Indian village on the north bank, that had been abandoned before it had been [fully] built. There was really some excellent native work about it, and I naturally asked its history, expecting to hear of some devastating epidemic, of course, but fortunately my theory proved incorrect. Its abandonment had been determined upon some five or six years before, when the builders saw the swelling "boom" at Juneau, with the Treadwell Mine alongside developing rapidly, and the attendant work which all this activity promised them if they moved alongside; which they accordingly did.

Nearly opposite Klame-quah there is a pronounced valley enclosing a large river, according to native accounts.

About ten o'clock the weather cleared enough to allow some photographs to be taken of the larger glaciers on the two sides.

The third glacier came in from the south side, and was the only important one from that direction.[7] By noon we were abreast of its front, which was from two to three miles away across the valley.

---

6. At the time of Schwatka's visit the two arms of Twin Glacier were joined in a single lobe at the base. With subsequent retreat of the glacier, this lobe is now occupied by Twin Glacier Lake. (See USC& GS 1896, Map No. 22).

7. Wright Glacier, the only glacier reaching the valley floor on the south site of Taku River, was so named by Hayes for his professor at Oberlin College, George Frederick Wright, 1831–1921. Wright (1887a) spent much of the summer of 1886 in Glacier Bay, making detailed field observations of Muir Glacier. While en route from Oberlin College he made further glaciological observations at a number of locations in southeast Alaska and the Pacific northwest (Wright 1887b).

Low islands covered with poplar or "cottonwood"[8] were now numerous in the course of the stream, and on one of them we spent a half-
hour getting our lunch.

During our short stay Robert prospected for salmon with the Indian
spear or hook of this region, but he was not rewarded. This fishing
implement of these Indians is a very simple and effective affair. It is
nothing more than a huge barbless hook of iron or steel fastened into
the end of a slim ten or twelve-foot pole, of some tough elastic wood.
Standing on the bank of a swift stream, generally where the curving
current is cutting into the shore, the native fisherman reaches far out
into the muddy water with this hook, point downward, and slowly drags
it along the bottom toward him. If a fish is felt by this prospecting, a
sudden jerk is given to impale it upon the hook, which, if successful, the
fisherman must display all the activity of a small boy with a pin-hook
who has also fastened to a fish, for there is no barb on it to retain the
salmon. In this way these natives, so they told me, catch all their salmon—
a not inconsiderable share of their diet; while hand-nets, native seines,
fish weirs and other Indian fish-traps are comparatively seldom used.

During that afternoon we passed several deserted Indian villages of
more or less permanence, from a roofless house of huge logs to the
temporary abode of bark slabs. All were fishing "ranches," the
occupants being farther up the river, hunting bears and mountain goats.
Later in the year they return, repair the ranches, and go to fishing. At
every village or house one will see a well-worn path of a few hundred
yards at least, on either side, closely hugging the river-bank. It is made
by the fishermen of the village prospecting for salmon.

It was nearly two o'clock when we came to the first narrow place in
the river, whose swiftness here faintly suggested rapids, although hardly
worthy of that name.[9] We got both sails fairly set to a strong wind, and
by dint of almost ferocious paddling and Wagnerian war-whoops went
through like a ricocheting projectile. An island, with the first stone banks
we had encountered so far, formed the obstruction in the width; yet it
soon widened and the doctor pronounced the Takou as yet fully equal
to the Tennessee at Chattanooga in width. The Takou is not very deep,
like all swift streams, and in many places we materially assisted the sails
by vigorous poling at the sides.

---

8. Black cottonwood (*Populus trichocarpa* Torr. & Gray). This is a coastal species,
not found in the interior. (Viereck & Little 1986, pp. 74–76).

9. The "rapids" referred to are located at Canyon Island.

About seven o'clock we went into camp on a little island, but everything was so wet and damp, from a recent overflow, that we crossed over, with some trouble, only to find the other side just as bad. However, we had had a splendid breeze all day, and a fine day's work to show for it, some twenty-five miles of travel from the mouth. There was many a weary week between us and the next twenty-five-mile journey; but dense ignorance was big bliss this time, so we curled up in the muddy leaves and went to sleep.

Shortly before noon on the 27th the Indians descried mountain-goats on the high southern cliffs, but they were too tired with the hard poling against the swift current to even suggest going after them. These Alpine animals are very numerous throughout the Alaskan mountains and those of the British Northwest Territory, and I have never yet pierced a range (I have cut through seven altogether there) without seeing them in greater or less abundance.

Ocher Bluff,[10] on the north side, passed in the morning, was so named from its hue, probably caused by prominent iron stains. It was an unusually warm color here by contrast with the cool greens of the river valley, and the decidedly cooler whites of the mountain tops covered with snow and ice.

The river was now a scant mile in width, and spreading all around cottonwood islands, from bluff to bluff. In many places the big canoe quivered like a high-pressure steamer, when the Indians were poling against the swift current. This shaking of the boat, or rather of the poles against the boat's side, is very trying to the nerves of the polemen, so they claim, and more exhausting than severer labor of a purely muscular character. Worst of all, the wind had died down almost completely, and we now fully appreciated how much it had helped us the day before, when the current was not so swift.

The doctor's geological researches suffered especially by the condition of things. It was only at the great convex bends where the river would expose the rocks in places, and these the Indians religiously avoided in their fight against the current; and then, as if to add "insult to injury," the best available evening camp was always on an island just as exasperating as on the water.

Late in the afternoon mountain-goats were numerous all around us, in groups of from two to four, on an average, and in general from 1,000 to 2,000 feet above the valley level; so when the work got so hard

---

10. Ocher Bluff is shown on modern maps as Yellow Bluff.

poling and paddling that even climbing the crags after goats was comparatively easy, the Indians desired to exchange, and we pulled up into a small back-water bayou, on the north side, to hunt these Alpine antelopes. The nearest and most available group was a nanny-goat and two kids, and they certainly ought to have seen us, although we did keep reasonably quiet, for savages. Still, from what I saw of them that day, they are stupid game, and the greatest excitement in hunting them lies more in the dangerous country they make their habitat than in any extra wise precautions necessary in stalking them. The group of goats selected was from 1,000 to 1,200 feet above us, and two Indians started from the mountain's base to climb for them, their arms consisting of a rifled musket and my Winchester, 40–82 magazine gun, as I felt too tired watching the Indians pole and paddle all afternoon to attempt it myself. From the time the two Indians left us, nothing was seen of either by the remainder of the party, who sat in the canoe as if it were a private box in the theatre, until nearly a half-hour had elapsed, when one shot was heard, followed almost instantly by the other. The she-goat fell at once, and so steep was the mountain-side that her death-struggles carried her 400 to 500 feet "head over heels" down the precipitous slope to a small shelf, where the body lodged. The little kids ran around in a dazed, stupid way, until two or three more shots were fired ineffectually, when they scampered off as grace-ful as gazelles. Their meat was probably worth eating but that of their dam was stronger than a badger's. The Indians slowly made their way down to the dead goat, and gave it another fling that helped it down a few hundred feet, and so it arrived by stages or on the installment plan, and, as usual with that plan, in a very dilapidated condition. When the animal reached us it had one horn missing, and I do not think there was an unbroken big bone in its body. They told me this was the orthodox way of bringing a mountain goat's carcass into camp from a neighbor-ing height, and they thought it made the meat tender. If this particular goat was any tougher when it started than when it arrived, and it was a fair average for the species, then I am willing to believe the stories of their jumping hundreds of feet, turning double somersaults, lighting on their heads, etc., etc. that I formerly doubted.

So much time had been occupied in the hunt that we went into camp on the shores of the bayou; or, at least, the Indians did, for the smell of the butchered goat drove the white men away, and they slept a few rods distant. I never saw men on a trip care so little for fresh meat.

This day but six miles were made, yet at the close the boatmen were very tired.

The next morning a fair breeze lured us into hurrying away, but it rather slackened as we started, and, tired out poling and paddling by eleven in the forenoon, the canoe was stopped on a driftwood bar for lunch, and to await the freshening of the afternoon breeze, which the natives assured me could be usually depended upon an hour or two after the sun had passed the noon-mark. A meridional observation on the sun for latitude was taken advantage of during this delay; also a few photographs. We failed to realize on a good wind for the afternoon, yet it did not desert us wholly.

A few river valleys were passed during the day to right and left, the most important containing the Salmon River of the Indians, coming in from the east. Just beyond, we went into camp on an island densely wooded with poplars. Here fresh black-bear tracks were in abundance, but it was the only part of the animal we saw. Seven miles to our credit were all we could count, as we turned in for the night and listened to the avalanches every now and then.

One of the surprising things to me was the kittenish playfulness of the Indians after camping, even though a very hard day's work had fallen to their lot. Many of them wrestled and played "tag," running all over a 160-acre island, while one amused himself cutting down a number of large poplar-trees just to see the fun that would be created by others, who were resting under the shade, scampering to get out of the way. All of this was the most boisterous "horse" play I ever saw to be following a day of hard labor. On my three Alaskan expeditions in which I have used these T'linkit Indians, this merrymaking proclivity has always come to the fore, and oftentimes, as in this case, when least expected.

From a six or eight-foot length of one of the cottonwood-trees, an Indian carved a totem-pole of the usual hideous insignia, and we set it up with much savage ceremony, while waiting for the breeze to rise, the morning of the 29th. I suppose it is standing there yet, for it was sunk a foot into the ground, but the red pasteboard eyes have no doubt melted, and the japanned tin ears fallen away. I pity the child that comes suddenly on it around the bend of the river.

It was now apparent that the country ahead was getting less mountainous, and that we were ascending to an interior plateau, of which this coast range was the outer rim.

The wind did come up about ten, sure enough, and we got away under full sail; but we needed every puff of it, for the river was now very swift and strong. On some of the swift shallow places we could hear the rolling pebbles beat against the canoe bottom almost like hail on a tin roof. In some places it was seemingly a fight for life, in the ten to twelve-mile current, to avoid being swept down some side channel, swift, shallow and bristling with drift-timber. Oftentimes an hour's terrible work would show only a few yards gained. Backward and forward they would swing across the roaring channels to get good poling bottoms, each time losing ground that would almost make them break their backs to regain and add a little to. It was impossible to "track" in many places, and it was not every point where the canoe could be crossed on account of the dangers I have already mentioned. Once we were caught directly against the trunk of a huge drift-log in mid-stream, over which the swift water boiled in a perfect cataract. Why we were not swept under it or broken in two even the Indians could not explain, and like all inexplicable things falling to either white or savage, attributed it to some sort of superstitious luck. We could not budge an inch, so tightly was the long canoe clamped to its undesirable moorings, but we were finally liberated in a way the very least expected; the log itself broke, and this too, where it was a foot or two in diameter.

This nearness to an annihilating accident did not stagger the natives in the least but they buckled down harder than ever, poling and paddling like pirates, and yelling like panthers. There was not enough room along the sides for all to pole effectively, and as the doctor's time and attention was drawn to his topographic work a considerable part of the time, while if I stood up I threw the hydrostatic equilibrium of the whole expedition into an epileptic fit, we two were excused; but Russell, a first-class river-man, held his own with the others, and perspired his full share, of the bread-earning. In the use of the paddles, however, we came to the front, and broke our full proportion.

Late in the afternoon of that day we came to a well-built but incomplete house, with several ramshackle shanties around it that the Indians named *Ka-ko-quick*. There is a misty story that the main edifice had been built by an old French-Canadian *voyageur* and trapper, Budreaux by name, who intended to make this a trading station, probably. There was some misunderstanding with the natives, and the partnership was dissolved, since which time Budreauxville has not prospered, Juneau being nearer and better prices paid than usual with frontier traders.

Just beyond is the only cañon on the river so far, and it is hardly deserving the title. Here the whole river passes through a rocky gorge, with not very high banks, and some 125 yards in width. As the current did not increase its swiftness greatly, if at all, I inferred it was of unusual depth. Anyway, poling was out of the question, and it required strategic tracking to get the great canoe through.

A few hundred yards beyond this contraction the river forks into its two main branches, the more important being the South Fork [Inklin River], according to prospectors and Indians both. It was up the North Fork [Nakina River,] however, that we turned about a mile, and, as usual, camped on a gravel bar, after eight miles of travel that in labor represented nearer eighty.

At the confluence of the two main forks was a deserted Indian house, surrounded by a picturesque semicircle of graves, themselves diminutive houses. The whole river, however, bristles with picturesque graveyards and deserted or half-destroyed buildings, until one is forced to acknowledge that it is a solemn and melancholy old stream.

THIRD LETTER

I have often spoken of the universal, almost monotonous, green of the valleys; and yet, in places, the various shades were often prettily contrasted in the peculiar blending of the deciduous and evergreen timber. The light-green poplars, or "cottonwoods," were massed in the low bottoms, but they often threw out flankers up the mountainsides, until they slowly disappeared as vivid green specks high up the slopes among the dark-green conifers that, in turn, crawled down into the flats as isolated specimens here and there, and seemed almost dismally dark by comparison. Where timber fires—no doubt the work of wandering native hunters—had swept the slopes, the conifers stood out blackened and bare, while the green-hued poplars lived, making the contrast even yet more vivid.

The night of the 29th–30th of May gave us a light frost; but as nearly all vegetation seemed fairly well along, it demonstrated, I assumed, its very hardy character. A slight wind helped us as we started the morning of the 30th, but it was variable during the day. Despite the cold, the river rose some three inches, and I find in my journal the mournful prediction that "If the Indians keep camping lower and the river keeps rising, the camps and the cold water will soon meet." By noon we had made only a mile and a half, the morning's camp in sight and the polemen about exhausted.

Here a well-marked stream comes in from the west, with a broad valley.

I could not help thinking that the wind which we had had so far was the same old southern summer breeze that sailed our cumbersome raft over the half-dozen lakes on the Upper Yukon's course in 1883, and that it is a pretty fairly constant factor in these parts in the spring and summer seasons.

Early in the evening, tired out, we camped, not even attempting to improve on a coarse gravel bar for a bed-place by going a little farther. However, the big canoe was unloaded and beached, for the recent strains had started cracks, and she was lined with mud and leaves.

That evening I heard a ruffed grouse drumming in the woods near camp, from landing time until I dropped asleep, about half-past ten. So far we had seen none of these birds, but along the river had met swallows, kingfishers, a dozen kinds of ducks, cormorants, gulls, terns, snipe, robins, crows and eagles.

The many beautiful flowers seen to-day recalled the fact that it was Decoration Day at home—one of the most beautiful of our holidays; but it had been anything but a holiday to us.

The night of the 30th–31st the river rose a scant inch, but Robert insisted that the stream was now so swollen that unless we got a strong up-stream wind we could go no farther, although it was but a very few miles to the head of canoe navigation, where river transportation ceased and all effects would have to be carried on our packers' backs.

It was just after the noon hour that we got away with enough wind to spread our sails, but these had to be always aided by the poles, paddles or "tracking" from the shore. At one swift swirl around a point of driftwood, we were poling, paddling, sailing, and tracking all at the same time. Tracking was the most unsatisfactory, as being the slowest and most uncomfortable floundering through the wet and mud, while the nearness to the shore necessary to do it nearly always took the wind out of our sails, however strong it might be blowing on the open river. Where the swift river cut deep into the overhanging black-alders the tracking was exasperatingly slow and tiresome.

By 4:30 we had sighted the houses at the head of canoe navigation, a couple of miles away, but it took three hours hard work to reach them.

The country was now much flatter and more open, with the timber growth extending over the tops of most of the hills.

At 5:30 we made a desperate yet brilliant fight through high rolling rapids to attain a certain point, which we secured by the very epidermis on the skin of our teeth. It settled the day's success, however, and by 7:30 we were camped at the head of canoe navigation, and all river transportation behind us until we should float down the next one.

Many fresh bear-tracks were seen that day, showing how numerous this animal was in this locality, an inference which the Indians amply confirmed.

Native caches, dwellings, and grave houses at the village of Nakana, at the head of navigation on Nakina River. Canoe Landing, May 31, 1891. (U.S. Geological Survey photograph by C. W. Hayes) [Hayes #296]

This day we had made but three and a half miles, as far as the banks of the river were points of reference, but taking the water itself, we had probably pulled and pushed our way through twenty times that length. At least the Indians thought so, for they came and begged to remain over and rest, before starting on the still harder journey of carrying 200 pounds on their backs over the mountain passes.

June 1st broke as a beautiful day on the Indian village of Nahk-ah-náhn, as the natives called this picturesquely situated place at the head of navigation on the Takou.[11] A fine brisk breeze bowled up the river in a most exasperating way, now that we had no use for it. The day's delay was taken advantage of to determine latitude and longitude, rate and error of chronometer, etc., although a noon observation was nearly spoiled by the overspreading clouds that were just then coming up.

---

11. The site of Nahk-ah-náhn (Nakana) is now known locally as Canoe Landing. Here the Nakina and Sloko rivers join.

We were now getting far enough inland to leave behind the greater moisture of the immediate Pacific Coast, and the Indians assured me that it was dry here when the sea-board was usually drenched with showers. However, this was only a half-way point, so to speak, and farther inland it was yet much drier.

From now on, we might also expect to see more game, they also claimed, being now in the land of the caribou, moose, mountain goats, mountain sheep, black, brown and grizzly bears (if there is really any specific difference between the latter), wolves and rabbits. Deer, alone, are not found on the main-land.

A crimson-throated humming-bird was seen near camp, sucking among the wild blossoms.

A strange Takou Indian came into camp, spreading the cheerful report that the snow was from ten to fifteen feet deep some twenty miles ahead on the trail. I thought at the time that he was looking for a vacancy in our guide department, since none of our packers knew any too much about the trail, and a little later I felt satisfied of it.

A day's rest, with Indian allies, is only so in a physical sense, and the chances are twelve out of a dozen that before it is over the mental worry and irritation will more than compensate for it. This day was no exception. First, they wanted me to employ a guide, as they knew so little about the trail; and secondly—but I will lead up to thirty-fifthly, as sure as a sermon, if I start out to do the subject justice. As usual, I yielded a point or two, and put my foot down on the most. The guide business was settled by Robert's volunteering to go, as he knew the trail well, and would otherwise have been idly waiting at Nahk-ah-náhn for the packers to return, as he had agreed to do in his contract with me.

The doctor was about nearly all day "geologizing" among the hills. He had seen a great many ruffed grouse (the "pheasants" of Pacific Coast parlance) and many signs of large game. He had reached 3,100 feet above the river, and had had a fine view of the surrounding country. It was much more Alpine in character toward the coast than farther inland, where a great plateau or gently rolling country seemed to exist.

Now that the Indians will undoubtedly enter closer into the descriptions of the adventures, it may be well to mention them here more in detail; so I give their names below, both native and Anglicized, and the tribes to which they belonged.

Native Name	Americanized Name	Tribe
Kook-sáh'k	Robert	Takou*
Shak-qua-tãh	Sam	Sitka*
Kah-eé	John	Kootznahoo*
Tah-wõõt'z	Barney	Taltan
Skeet-lah-káh	Edward	Chilkat*
Koo-nagh-ka-sáh	Jim	Takou*
Kool-teén	Paddy	Takou*

All marked with an * as belonging to the great T'linkit nation, speaking that language. The Taltan are an interior tribe on the Taltan River near the Cassiar Mines of British Columbia.

Skeet-lah-káh, a great raw-boned Chilkat young man, had served me as a packer when a boy of but fourteen years of age, in 1883. He was the son of the Chilkat chief, Shot-rich, who assisted me very materially in my expedition of that year. Even then this Indian boy carried sixty-eight pounds over a forty-mile mountain trail, and this time was clearly the strongest and most enduring of all the packers. His Americanized name of "Edward" was a recent one, I thought, probably even made for the occasion, as he seemed to take strangely to it, while all others got theirs from the miners, trappers or traders. Sitka "Sam" had attended mission school in his younger days, and knew the English language well enough for "rough and tumble" interpretation, but he was not reliable beyond that. "Paddy" had also acquired a small share of it among the miners of the Cassiar regions; while, in fact, all of them understood it better than they pretended.

It will be remembered that the Indian packers would have to "double" their loads of 200 pounds over the trail; that is, after taking 100 pounds, the usual packing load, to its destination, or part of the way, they would have to return for the second load. In order to keep the effects together for safety, especially at night, it was thought that five miles a day over an average mountain trail would be a fair day's work. This would give ten miles of packing 100 pounds, and five miles returning, which, by comparison, would be equivalent to a rest, so they argued themselves. It might be well to state here, even in anticipation, that at first they generally carried the first loads into camp (four to six miles), and returned for the second, but soon adopted the plan of carrying each

The expedition packers ready to start on the trail from Nakana to Teslin Lake, June, 1891. (U.S. Geological Survey photograph by C. W. Hayes) [Hayes #295]

installment by easy stages of about a mile before returning. This gave them many shorter rests, while the rear loads were always nearer camp, wherever its suitability or their tired condition suggested it should be made. The white men took loads of from twenty to thirty pounds, being mostly clothing, photographic cameras, scientific instruments or other material directly useful upon the day's journey on the trail, over which they passed but once.

Early the next morning, well before six o'clock, in fact, the Indian packers started with their first loads, and we were almost dumbfounded when, at half-past seven, they returned to us at the camp for the second, stating that they had made some four miles on the trail. At this rate, we could easily make ten miles a day, or the whole portage in a week. Some of the more enthusiastic had promised us ten or twelve miles, if the trail was not too severe; but I knew the limits of Chilkat endurance from former experience, and I also knew that the Chilkat packers were better ones than the party we had with us. My opinions were confirmed when, later, the four miles dwindled down to a scant two; still, I thought to myself, as I wiped the perspiration from my face, that I might have deemed it a good half-dozen if I had carried 100 pounds over it, instead of a camera and a tin cup.

We got away with the second packs at nine, the white men accompanying the party this time. Everywhere on the trail we saw the "jack pine" of the frontiersmen—the *Pinus contorta,* I believe, of botanists.[12] It is a flexible conifer, if there ever was one, and every now and then we would come to a small "jack," of, say, two branches, that would be tied in a hard knot by the Indians, or two trees tied together, to mark the trail, as Robert explained. This had been done years before when the trail was new, but the pines had grown right along, just the same, and these knots in the bodies of the now large trees looked curious enough to stump the curiosity of any one not acquainted with the explanation. Stunted crabapple bushes in bloom and fields of juniper bush greeted us on all sides.[13]

We were now ascending an Alpine branch of the North Fork of the Takou, a stream too large to ford or cross, so we kept well on one side, the northern. Where we found the first packs there was also camped a Takou family from the interior, bringing out furs to trade at Juneau. It was this camp that had produced our would-be guide Kok-táche (Billy) by name. They had some furs rolled up in water-proof skins, but we did not have time to look them over carefully. It may be well to mention here, probably, that we met several of these fur-trading parties of Takous coming out, this trail being mostly used for such commerce.

The principal furs of southeastern Alaska are, in their order, black, brown or grizzly bear, mink, beaver, otter, land (sweetwater) and sea, martin, lynx, wolverine, silver cross and red fox, and, occasionally, white and blue fox from the lower Yukon.

A prominent furrier of this region told me of his first experiences, many years ago, trading with these people. He once priced a fine silver-gray fox skin, for which he would have gladly given thirty or forty dollars, or even fifty at a pinch; but to his astonishment, the Indian

---

12. The inland variety of lodgepole pine (*Pinus contorta* var. *latifolia* Engelm.). (Viereck & Little 1986, p. 48).

13. Oregon or western crabapple (*Malus diversifolia* (Bong.) Roem.). (Viereck & Little 1986, pp. 162–165); Common juniper (*Juniperus communis* L.). (Viereck & Little 1986, pp. 68–69).

trapper, in broken English, asked only sixteen. The judicious Indian trader, so it is said, never "snaps" at any bargain; so my friend began to argue, ostensibly to secure a better price, but the Indian was firm. The "dickering" was long drawn out, for time is of no value in these trades, until my friend finally laid down the sixteen dollars, having obtained a slight concession in regard to something else, adroitly brought to the fore. He was equally astonished to find the Indian refuse it, and an explanation led to the fact that the native had meant all the time "six tens"—or sixty—which he proceeded to show on his hands, each extension of the fingers meaning ten, their savage unit of figures. I have seen it argued somewhere, by a professor of political economy, that if the United States had adopted the golden eagle (ten dollars) as a unit, instead of the dollar, it would have had a corresponding effect on its capacity for wealth; as notice France, with its franc (twenty cents), or Germany with its mark (twenty-four cents), can never become the money-center of the world against Great Britain with its pound (nearly five dollars). True it is, these Alaskan Indian traders are very well off, for savages; but whether due to their unit of ten fingers, or their shrewd bargaining, I will leave to others to work out.

All day we were crossing beautiful mountain rills with water cold as ice, so thirst was out of the question despite its being a warm, dusty day. All were inclined to perspire more or less freely, but as usual with one wearing glasses, I caught more than my share of the discomforts. Without them I could not distinctly see distant objects, while half the time they would cloud up from the perspiration until they were as opaque as ground glass. Of course, when most needed over rough, rocky ground or fallen timber, etc., then the glasses were the worst, as a natural consequence of the cause of their blurring. In the Arctic the frozen or freezing breath took the place of the perspiration, and, in general, the near-sighted explorer is short-sighted indeed for going on a rough trip.

At 2:45 we got into camp by a mountain rill that looked like buttermilk as it dashed down the steep mountain-side. It was some four miles from Camp 7, at head of canoe navigation, and we expected to make farther, but the Indians did not get up with the second packs until 7:30, too late to make any distance worth the while. This Camp 8 was where a great dike ran across the stream through which the latter cut by a most picturesque cañon, bordered by "nubbins" or precipitous hilly summits, where the hard dikes had not disintegrated, the two nubbins, like the posts of a gate-way, being visible for miles up and down the Alpine valley. Back of either, a pack-trail, wagon-road or even railway could be made.

That night I learned a new "wrinkle" on roughing it. I had selected a beautifully prepared spot in a slight hollow full of leaves for making down my bed. It was cut off just right at the end by a fallen log, two-thirds buried in the turf, that formed an excellent pillow. The next morning the Indians were up by three to get an early start, and they began cutting firewood from the part of my tree that projected from the slope. There were 9,472 other logs in the immediate vicinity, all of them dry, solid and reeking with pitch, while mine was water-soaked, rotten and heavy; but they kept beating a tattoo on it with axes until five o'clock, when they got away with the first packs, all of them stumbling over the only part of the trunk they had not yet cut up.

There were now many limestone bowlders and pebbles in drift, showing large deposits somewhere farther up the stream. Small bird-hawks and wood-peckers were seen near this camp—a rarity, so far, on the trip.

It was not until the middle of the afternoon that the Indians returned for the second packs, when we all got away. For the first mile or two the trail was strewn with fallen timber, but as soon as we got out of this the travelling was very much better. The Indians are quite careless with their fires; and while their camps do comparatively little harm, the signal smokes they often send up when travelling to convey intelligence to others far away, are prolific sources of great timber-fires, sweeping considerable areas. In a few years a fierce storm prostrates the burnt standing timber, and a walk through the woods is then a funereal hurdle-race. How the packers got along so well over this labyrinth of logs surprised me more than my own constant falling off of them. I had reserved my hob-nailed shoes for any serious mountain work that might occur, but after slipping off of half the logs that day, skimming over the slick pine-needles in the path, sliding all over the wet moss of the hillsides, and essentially skating the full five miles into camp on the smoothest-bottomed shoes I ever wore in my life, I came to the conclusion that the serious work had arrived.

Again to-day, we saw interior Takou traders making for Juneau with furs, this being about the end of their season. Bear-skins secured in the spring are the best. From the end of May until January they are worthless. When the skin side of any fur-bearing animal is dark colored, it is safe to assume that the fur is poor. The lighter the color, the better the pelt. When a bear's hair shows through to the flesh side, it is a poor

Native packers in front of house, Nakina River, June 3, 1891. Note the two folding canoes leaning against the end of the house. (U.S. Geological Survey photograph by C. W. Hayes) [Hayes #327]

skin. It must not be forgotten by the admirer of these robes that many Indians, in moments of forgetfulness, will glue loose hair on long-killed animals, just to brighten up the pelts a little bit.

That day, the 3rd, we camped at the deserted Indian village of Ah-kah-tée, near which there is a native grave-yard overlooking the river. That was one of the toughest days on the trip, the Indians being on the trail from four in the morning till eight at night. Of course, they took long and frequent rests, but I think they were needed ones. That evening several complained of strained tendons, and one of them adopted the heroic treatment of standing half an hour nearly knee-deep in the ice-water of the mountain stream. Fortunately, the weather was cool and cloudy, or the results might have even been worse. Robert secured a grouse on the trail, the first bird of the trip, and this, with a part of a salmon, made up our evening meal.

Next day it was 8 before the packers got away, and we all got started by 10:30. It could be noticed that the Indians were getting a little sore where the pack-straps cut, and some of them were seen to tear up great, broad strips of blanketing to make new shoulder-straps for packing. We seemed to cross an unusual number of pretty mountain rills that day, while often a glimpse into the valley would reveal an occasional deserted Indian house. Lunch time saw us two miles on the trail, coming up to the first packs at the village of Klick-nóo (deserted), just above which there is a beautiful waterfall on the river.

In this vicinity are many prominent dome-like hills, one of them being quite conspicuous, bearing to the south. The doctor said that it was marble, reasoning from a similar one near by. That night we camped at a place the Indians call "the pole-bridge" where a number of pine poles had been thrown from either shore to meet a high rock in the center of a foaming cataract.[14] These pole-bridges are not uncommon on trails, and as the trunks are usually free of bark, and slippery with the water that has dashed over them, they are not the pleasantest way possible of crossing a dangerous stream. That evening the packers did not arrive in a body, as usual, but were strung out, according to their strength and endurance, from five until seven o'clock. It was getting clear that some of them would fail me, if the road ahead got any worse, or the slightest addition should happen to the labor.

Next day, however, Nature gave them a rest in a way not greatly expected; in fact, by a drenching rain-storm that night and the next forenoon. So far we had all bivouacked in the open, but that night we got a good wetting as a result, and the tent went up next morning for the first time on the trip. At noon, the weather clearing, the Indians decided to carry forward a load, such as the canvas boats, etc., that could be left in the open, and this they did, getting back very late.

Robert stayed over in camp until the middle of the afternoon, when he started back for the head of canoe navigation, assuring me that he would reach it that night, and overtake us next day on the trail. He evidently thought we were making slow progress, and that he ought to return for more provisions.

The doctor spent the afternoon looking into the geology of an adjacent high range which he pronounced limestone, while Russell did some hunting, with unrewarded efforts.

---

14. The pole bridge was across the Silver Salmon River near its junction with the Nakina River.

T he half-day's rest at the pole-bridge camp also gave them another half day on June 6th, caused by the packers losing the
trail during Robert's short absence. He had explained the split-
ting of the trail to them thoroughly (so he claimed) before leaving us,
but in some way they managed to take the wrong one.

A few miles beyond the pole-bridge this great trail splits into its two
main branches; the northern one, which we should *not* take, being the
shorter, and terminating in my old explorations of 1883 at a lake which
the local Indians (Tahk-heesh) call the Takou (or Tahk-óo), because it
leads to the Takou country, and which the Takous, as well as all other
T'linkits, designate Tahk-heesh, for reasons easily inferred.[15] This would
make the proper name the Takou; but the T'linkit influence so pre-
dominates, even in adjacent countries, that their names readily take the
lead, especially with those persons employing the latter Indians.

Before the Civil War closed, the Western Union Telegraph Company
tried to connect its lines with those of Europe *via* Bering Straits, and
much exploration was done with this object in view. This Takou trail to
the lake mentioned is the one, I think, traversed by Mr. Byrne, from
conflicting reports I have seen. His report, I understand, amounted to
the practical fact that it was available for a telegraph line, but I have
never heard of any other results.[16] None of the Takou Indians, even the

---

15. The northern trail led up the Silver Salmon River, crossing the divide to Pike
River and on to Graham Inlet and Taku Arm of Tagish Lake by way of Atlin Lake and
Atlin River.

16. Michael Byrne (or Byrnes), a well-known miner in the Cassiar region, explored
this area, reaching Tagish Lake ("Lake Takou") before being recalled by his employer,
the Western Union Telegraph Company. The geographic information he gathered was
incorporated in Dall's map (Dall, 1870).

older ones, who had spent the greater part of their lives on this trail, remember anything about it; and yet it undoubtedly took place. This northern trail is now very seldom travelled by the Indians, as the fur-bearing country it taps is given up wholly to the Chilkats, who reach it by way of the trail known as the Chilkoot, through the Perrier Pass.

The weather was yet heavy and threatening, the morning of the 6th, as we got away. The doctor and Russell had been obliged to repair the pole-bridge the evening before, the rains having brought down heavy enough driftwood to tear it out badly on one side.

It was a good morning for grouse on the trail, and a number were secured with the Winchester shot-gun by members of the party. These birds were very dark-colored, darker than any I have ever seen in lower latitudes. In fact, they seem to get darker the higher the altitude, until, with a suddenness quite marked, the pure white ptarmigan, or Arctic grouse, is encountered at and above the timber limit.

The way we came to miss the north-eastern trail was that the junction of the two was on a wide stretch of broken, angular gravel, down which an avalanche, carrying trees and Alpine *debris* of all kinds, had recently moved. Here the proper trail to take led sharply up the mountain slope, the northern trail being much the plainer of the two.

These numerous gravel slides, or gravel cones, on the mountain-sides were curious affairs, exasperating to travel over, but interesting to study. It was of an uniform, angular character, as if it had come from a rock-breaker and intended for railway ballasting or the macadamizing of a road. Here was enough to have ballasted the whole length of the Canadian Pacific Railway, the longest road in the world, and plenty to spare for constructing all the county roads that could be made in Yavapai county, Arizona, the largest county in the United States. Yet the doctor and Russell assured me they had seen very much more of it on the other slopes of the mountain the day before. If a railroad ever cuts through this country—which I think not unlikely, for reasons I will give in a later article—it certainly will be well supplied with this necessary and usually expensive material of construction.

At 9:30 we passed a huge split bowlder, buried in the brush and timber on the trail, that had painted on one of its flat sides a number of Indian totems and hieroglyphics. It was on the Lake Tahk-heesh trail, over which we had to return later, and not having the packers with us at the time, got no explanation of it. It was also in too secluded a place to get a photograph [on] such a dark, gloomy day. The work seemed very old, and was, no doubt, purely native in all its meanings, and was the only thing of the kind I have ever met in the T'linkit country anywhere.

As we rapidly ascended the mountain slope by a very well defined trail, after the gravel cones were left behind, we had the best view of the country that the route had yet presented. Between the two trails lay a great, bald mountain, with rounded top, and on this top were two conspicuous black objects, looking not unlike two houses, and these could be plainly seen for miles on either side. I think they were huge, erratic bowlders, even at this high altitude. I named this butte the "Trail Splitter," from its position with regard to the two trails already described.[17]

That night we camped on the high mountain slope, 1,350 feet above the river, as shown by the aneroids, and where it was not easy work to find enough available flat space to even make our beds so as not to roll down the hill. At this altitude the spruce was of the stunted, trailing kind, especially good for making a nice, soft bed on the ground.

In cutting open the grouses' gizzards for the evening meal, I noticed that they contained nothing but cottonwood leaves, which must be their standard diet here at this season.

In opening the bundle containing my bedding that evening, I noticed that the pack-straps of Edward, who carried my "outfit," had snugly rolled up in them nearly a pint of spruce gum-drops that he had evidently collected on the way, and showing that even the hardy son of the forest is occasionally addicted to the gum-chewing habit.

That afternoon, just as we were nearing our mountain camp, Robert came in sight, winding backward and forward up the steep zigzag trail, and soon passed us, thus redeeming his promise, which I had hardly expected, from the great distance he had had to travel. Sure enough, he had some thirty or forty pounds of provisions on his back, the main object of his return trip.

At this camp the Indians found a small pine squirrel—I think it was in the stunted spruce—and the way they turned out *en masse* with clubs, stones and other missiles after it, would lead one to believe that it was as important as a full-grown moose at least, or that they were on the verge of starvation. After an exciting chase of ten minutes or so—I have seen an equivalent amount of energy secure a dozen buffaloes or elks—

17. The north-eastern trail taken by Schwatka's party led up a side stream shown on recent maps as Katina Creek.

they secured the two-ounce affair, and proceeded to cook it over the camp-fire. I knew a native, at another time, to shoot one with the Winchester shot-gun, the cartridge weighing more than the carcass.

This camp (11) on the mountain-side gave us a very cold night for sleeping, as we had expected from the high elevation. That night, the first time for a week or over, we did not use our mosquito-nets, the light frost keeping them away.

There were a great many snow-banks on the trail that forenoon, which still led upward until the aneroids showed 2,652 feet above river level. These snow-banks are not very bad in the early forenoon, but later in the day, when it is warmer, they are very exasperating, and directly so according to their depth. Of course, the packers suffered the most; still it was not pleasant even for the others to sink in the wet snow up to the knees, at least, at every step, and occasionally the whole underpinning gave way, leaving one up to the arm-pits, if the ground was not struck sooner. Grouse were plentiful near these snow-banks— probably after water—and a few were killed.

The conifers of these high altitudes were covered with long, trailing moss, almost as black as jet. It hung in great festoons from the limbs, very similar, except in color, to the long, hanging, gray moss that trails from the oak-trees in Oregon and the live-oaks in Florida. It was so abundant on some of the spruce-trees that it had fairly smothered them, it seemed; or, at least, they were dead and leafless, with enough of this moss around them to form a black, gruesome shroud, so to speak. But these trees were fine ones for sending up signal smokes, and a match touched to the moss on one responded with dense volumes of smoke that must have been visible for many miles up and down the valleys.

Nearly all the trees and brush on the mountain-side were leaning down-hill, showing the effect of the winter's snow in bending them the direction it kept slipping. Many of the large trees were badly "barked" even ten to fifteen feet from the ground, no doubt caused by loosened bowlders, bounding down the slope in great jumps, striking the trunks this high while on their course.

Despite the melting snow through which we floundered, we found no drinking water anywhere, the gravelly and loose soil absorbing it all as rapidly as it was formed. As we descended the slope, however, the springs reappeared and by the time the valley was reached it was quite swampy in many places. Oftentimes these swampy places enclosed small, shallow lakes, and on the bottom of one we could see the white ice that had not yet melted.

Old moose signs were seen everywhere, but nothing fresh and not a hair of the animals themselves. Still we must now have been where they were numerous at certain seasons, year in and year out. These fine game animals have been seen on the Takou River, as far down as the Alaskan-British boundary; but these instances are rare in that direction, while as soon as the coast-range is fairly pierced and the interior plateau gained they may be looked for on any of the water-courses, where they love to browse on the deciduous trees even when the grass is sweet and abundant. In certain low localities probably the mosquitoes and gnats drive them to the higher, cooler altitudes. There is no other way to account for the numerous signs in some places and yet the entire absence of the animals themselves. As this ruminant, the only one, I believe, feeds upon evergreens as well as deciduous trees and shrubs, it would find plenty of this food at nearly the highest altitudes of the inland plateau.

Just before two we came to an open but brushy country, covered with a kind known to frontiersmen of this country as "moose-brush."[18] It is much harder to make one's way through when at all dense than any other kind I know of, and it requires a skilled backwoodsman, indeed, to follow a trail through it, unless very well travelled. In another hour the "moose-brush" had disappeared, and a beautiful park country stretched out ahead of us. It was on the summit between two forks of the Takou, and while pleasing to the eye as it broke on us, we found it wet and marshy enough under foot. Just on the summit is a large, shallow lake which is said to drain both ways; a not uncommon peculiarity of summit lakes. On its shores were camped a number of Indians, who were outward bound to trade furs at Juneau. Near them our own Indians had selected their camp, in order, they said, to trade for moccasins with the newcomers, their own being about worn out. This last statement was certainly true, yet I honestly confess that when I have savages in my employ, I detest their meeting, in any way, others of their kith and kind. They can sometimes enter into more mean, petty conspiracies in five minutes to worry one than Pinkerton's whole detective force

18. Hayes described "moose brush" as a species of alder, probably a subspecies of *Alnus viridis.* (Viereck & Little 1986, p. 140–142). (See also Viereck's preface to the 1986 edition of *Alaska Trees and Shrubs.*)

A group of interior Natives in camp along the trail. (U.S. Geological Survey photograph by C. W. Hayes) [Hayes #332]

could unravel in as many years. This time, however, they agreeably disappointed me.

For some time back, however, there had been complaints in anticipation of the food supply running low, and the subject was now getting to be quite a serious one. I refer to the Indians only. They had brought only about a week's supply; an amount which they can easily dispose of in half that time, and certain kinds were now running out. They were especially pathetic over sugar, a luxury that they seldom had, but which now, singularly enough, seemed indispensable. A couple of moose or caribou, which we had every right to expect, would have settled the question for probably the whole portage, but the chances in that direction seemed to be fading faster every day. So, when Robert asserted that it would yet take six days to reach the big lake where the packers' contract ended, they decided to send one member back to the head of canoe navigation after food stored there, while they would divide his packs among them and follow the trail until he overtook them. I agreed to feed them from my own supplies at cost, plus twenty-five cents a pound, which they were charging me for carrying it over the trail. This charge for provisions I exacted only as a check to the otherwise enormous Indian appetite, which would have bankrupted our commissary department by the time they left us at the lake, as I never intended to enforce it. It had the desired effect, and an amount of gratefulness that I had hardly expected from Indians when we afterward reached the big lake. They really gave two grunts, instead of the usual one, and I think I detected even cases of three among some of them. Paddy, a good-natured Takou of fat, squatty proportions, was selected for the return, and I curled up the corners of his mouth by giving him an extra pair of moccasins to make it in. He got away next morning before the morning had really got under headway—about two o'clock.

The Indians themselves got started late (6:30), and were back in less than two hours, having made only a short distance on the trail, Paddy's load appearing extra burdensome divided among them. They reported a large number of white grouse (ptarmigan) as having been seen in the open places.

We had not gone a mile on the trail before we came to a beautiful, bubbling spring, "boiling" two to three inches above the level in its center. Its water was ice-cold, and it was the only one of the kind that we saw on the whole trip. In the arid lands of western Kansas or New Mexico, it would have been worth more than a dozen gold mines paying twenty per cent assessment on every share. We saw nothing of the ptarmigans on the summit, and as soon as we began descending into the valley of the next fork of the Takou and entered the dense timber tracts, we gave up looking for them for the time. During the breeding season no birds are at their best, as every sportsman knows; and even the explorer, who is privileged more than any one to live off the country, should not kill more than is needed for fresh meat, unless a "pinch" justifies it; but during this time I think the white grouse is the superior of all in its family. It is usually the cock, by his coarse cackle, that attracts the hunter's attention, and more of them consequently fall than the hens, the kind easier to spare, besides being in better flesh.

About the middle of the forenoon we again came to a large party of fur-traders on their way to Juneau, and there was no mistaking their errand, for every one of them able to carry a pack was loaded with furs. There was one small Indian boy, of about twelve years of age, who had a pack of them nearly as big as himself. I tried to estimate the weight by testing the pack, and I think it was a fair forty pounds, while others thought it might be even fifty. The little fellow appeared proud of his burden as we inspected it, and even his father seemed happy, though married, for he had three wives. One of these squaws was carrying a very old woman—a great-great-grandmother, I should think, for she looked exactly like the Indian mummies of this country sometimes seen doubled up in half space. I should like to have had her history of this country, but she seemed well past the point of intelligence enough to give it in a coherent way. Alongside of it the stories of many frontiersmen would seem like a tame sort of civilization.

Just before noon we reached the next fork of the Takou at a second pole-bridge, it, too, being thrown over a foaming cataract.[19]

From this point on there were two trails, that rejoined before the big lake was reached. The usual one led over a high ridge of mountains that, at this time of year, was liable to be seriously obstructed with snow; while the other, although being the longer, simply kept to the lower levels and avoided the greater part of the snow. The reader may remember that I spoke of a party of prospectors from Juneau having made this trip to the big lake in the winter months, using sledges for their effects and without guides. They followed a native sledge-track in lieu of a guide, and it was near here that its course became so erratic, they thought, that they left it. They were eighty days going from the mouth of the Takou River to the big lake, often passing over the course five or six times between camps with their "bob" sledges, and averaging a little less than two miles a day. Russell had belonged to this adventurous party, and from his arduous acquaintance with it, knew considerable about the trail. Robert had returned with the party met at the summit lake, his labors, so far, being partly voluntary, and having explained the further trail thoroughly, as he thought, thus throwing more responsibility on Russell as the new guide.

By the way, I should have mentioned that we called the summit lake just mentioned "Trout Lake."[20] The Indians camped near it had one or two of these fishes, and so the white men soon had their tackle out and were after them. They certainly were the most apathetic trout I ever saw. As to rising to a fly, that was simply out of the question, even when it was dangled directly over their heads in the most alluring way. It wouldn't even frighten them away. Robert told us that the Takous caught them with spears, and he gave us some salmon roe to "pot-hunt" them with as bait; then our luck changed. By adroitly putting the bait about

---

19. Now known as Taysen Creek. A few miles upstream, Taysen Lake is shown on some recent maps as Paddy Lake.

20. Schwatka's "Trout Lake" is shown on recent maps as Katina Lake.

an inch or two in front of their noses, we put enough life into them to swim up and take it, so that we eventually landed nine of them, weighing from about two pounds down. Even their biting was feeble, and a number of those secured were thrown on the narrow beaver dam where we were standing, and the doctor had a "catch-as-catch-can" with them before they were strung on the willow switch which insured their safety.

All of the outgoing Indian fur-traders we had met so far advised us earnestly to take the lower but longer trail, and this we determined to do. There was but little trouble in finding it; but as the upper one was the plainer, sufficient signs had to be left so that Paddy would not go astray with the reserve provisions.

About the middle of the afternoon I shot a dusky grouse, and about the same time we lost the new trail for nearly half an hour. So long as the outgoing Indians kept fairly together there was little trouble in following their path; but every now and then they would scatter, in order to peel the inner bark from the young pine trees, and thus often confused us, until we learned to follow the barked trees at such places more than the foot-prints. They are very fond of this inner bark, it seems, although, to my taste, I must confess, it savored too strongly of raw pitch to be palatable.

That afternoon we crossed many peat-like bogs, or swamps, that, half a foot to a foot below the surface, were solid ice. Here the coarse grass often grew in bunches, so close together that the foot could hardly be put down between them, and, if placed on one, it turned over as readily as if it worked on a greased ball-and-socket joint. No professional wrestler could contend with them if he once started to fall. The black, swollen ankles at night showed plainly how it told on the heavily laden packers.

In the brush thickets we saw where hares had been browsing off the bark, as well as many other signs of them, but no animals whatever. The Indians have a curious theory that about every seven years these little animals have a period of maximum or minimum numbers. If this is so, we probably had struck the minimum period, not only of the rabbits, but of all other game that indulged in variable quantities, and that the signs hung over from the maximum. We had really game signs enough to have supplied a battalion by fair hunting, and to justify keeping out guards to prevent their running over the camp; but, outside of the gutta-percha goat and some wrought-iron grouse, we had not tasted fresh meat so far.

That night we could hear the roaring of the stream two miles behind, showing how quiet nature was, as well as the high stage of water in all the streams draining the locality.

From the trail that day, at one high point, we saw distant snow-clad mountains to the eastward, that the Indians said were beyond the big lake.

The next morning (June 9th) we had our first fog, the clouds lifting a little later.

Before starting the Indians sent up signal-smokes to guide Paddy on the trail. That forenoon we passed a number of small lakes exactly the color of black tea, the stain evidently coming from the steeping of the forest leaves. The third pole-bridge, over a most furious cascade, was passed just after. A beautiful rainbow spanned these falls, forming a fine subject for a painter's brush, could one have tarried there.

All that day we were plowing through moose-brush and floundering through marshy parks. The Indian trail seemed to prefer the bogs to the solid ground with its matted brush. Along the way, at every rest, signal smokes were sent up for Paddy's benefit, the packs, mostly covered with water-proof canvas, suffering several holes as the result of the falling sparks.

We now encountered a plant, a low, trailing kind, that afterward remained abundantly with us. It is known as Hudson Bay tea, and it is said that it makes an infusion similar to cheap tea flavored with nut-galls. The *voyageurs* of the H. B. [Hudson Bay] Co. used to use large quantities of it as a substitute for tea.[21]

That night we camped in a forest of dead, burnt timber, which, rotting as it stood, showed that the locality was not much given to severe gales.

Just as a couple of Indians were starting back to look up Paddy, that individual, covered with smiles, hove in sight. He had easily guided himself by the smokes and fires, and was loaded with "grub"—the Indian's only idea of the English language for food. That night the camp was a happy one.

After Paddy's return, we all thought that the food question was settled, outside of probably furnishing the Indians some to return by; but inside of two or three days it was gone again, and we were issuing from our own scant supplies. That night (June 9th–10th) there was another frost, and I had little doubt but that this cold weather had an appreciable effect on the rations devoured, where there was no definite amount issued.

Some of the swamps were now very wide, and one especially was so bad that we had to make our way very slowly over it on the zigzag course marked out by fallen logs. A couple of streams were so large that we had to cross them this way also.

There was much dead timber passed through the 10th, but it did not seriously interfere with us. The only trouble was the slight one that

21. This plant is also known as Labrador-tea (*Ledum groenlandicum* Oeder). (Viereck & Little 1986, pp. 207–209).

whenever grouse were flushed in it, they invariably flew long distances, and were seldom secured in comparison with the live timber. A hunter on an exploring trip, carrying a dozen necessary odds and ends besides his gun, is seldom prepared for an immediate wing-shot, and cannot follow up far-flying birds like a sportsman devoted wholly to the pursuit. I have spoken several times of the constant southern breeze, and more than intimated about our general course being in a northeastern direction. This combination alone would have kept away a great deal of the large game, had any fresh signs been seen to verify its presence. I am more inclined to think that as soon as this trail is passable and the fur-hunters begin using it, about all the large game leaves its vicinity. The only very fresh signs we saw were those of bears grubbing in the swamps for roots.

Just before noon we crossed the fourth pole-bridge, and, as usual, it was over a mountain torrent.

Mosquitoes had been getting worse almost daily, and were now so bad that the only comfortable place was in a dense smoke from a campfire of spruce or pitch-pine. Black gnats were also putting in an unwelcome appearance.

About the middle of the day on the 11th we reached the main North Fork of the Takou, the one we had left at the Indian village of Klicknóo.[22] It is here about twenty yards wide and very swift. Not a bite could be had by fishing in it with either bait or fly. After ascending it a mile or so we went into camp to await the packers, who did not get up with the second load until eight, being the best drive we had yet made, something over seven miles.

It was at this camp that the little black gnats were the worst we saw them on the trip. The doctor busied himself building smudge fires all around the camp, but as there was little wind to drive the smoke along, it ascended directly, and had only small effect on the millions of mosquitoes and gnats that buzzed around us. We clearly demonstrated that a strong, dense smoke will drive away mosquitoes that can be brought within its influence, but has no apparent effect on these gnats. In fact, they rather seemed to like it, and were the thickest in great dense bunches directly in the smoke, and from ten to twenty feet above the ground. There was nothing but seasoned spruce[23] near camp, and it gave but

---

22. This fork of the Nakina River is shown on recent maps as Hurricane Creek.

23. White spruce (*Picea glauca* (Moench) Voss). The tree is common throughout northern Canada and interior Alaska. (Viereck & Little 1986, pp. 52–54).

little smoke, as the Indians did not see it at all, but followed us wholly by the trail. It will not compare with green trees, as one would expect, for signal smokes.

This camp was where an old Indian stopping-place had been, as seen by the refuse skins, broken and abandoned snow-shoes, etc., etc., and it was undoubtedly the crossing-place on the new trail, as described to us; but at no point could we find the raft left by the last party of Indians who had told us of it. The high water had probably carried it away. We, however, set up one of our folding canvas boats and used it as a ferry next morning.

After the Indians crossed on the morning of the 12th, they spent an hour cutting logs for a raft on which to recross as they returned to Juneau later, as they would have no ax with them on that trip. Later, the white men crossed with the last loads, and after waiting a half-hour for it to dry, rolled up the folding boat and got away at nine. Another hour, and heavy thunder was heard near by, while at noon a brisk thunder-shower, with considerable hail, sprang up and lasted for two hours. Many of our Indians had never heard thunder before, and the vivid lightning and deafening peals of this long, fierce storm must have been rather a startling initiation. They said the great rumbling noise was caused by the flapping of the wings of a huge, dusky bird, and some of them positively asserted that they had seen it that day. Whether this is the superstition of the interior Indians, who encounter these storms frequently in the summer, or whether it was one suddenly made up by these novices, I could not find out. Practically, the Pacific coast of Alaska, as well as most of that on the Bering Sea, is unfrequented by such storms, but the interior plateau has its full share for such high northern latitudes.

These rain-storms always seemed prolific in dusky grouse, and we secured several that day. Russell, in shooting one, cut off an inch and three-fourths thick green spruce limb just beyond the bird, which fell with the *débris*, showing the great concentration and power of the Winchester 10-gauge shot-gun. This gun carries six shots in its chamber, but can be loaded as an individual breech-loader at any time that the hunter may want to change the character of his loads. Over twenty-five thousand miles in the field, exploring, hunting, scouting and marching has shown me that this rapid-fire shot-gun, coupled with a 40–82 Winchester rifle and a Smith & Wesson 40 or 45 caliber six-shooter, practically arms the bearer for all sorts of game and dangers to be met by fire-arms as thoroughly as it is possible for mechanism to do; beyond this it is a matter of sportsmanship or a contest between wood-craft, prairie-lore and general experience and intelligence.

That day's march was a fearful one through wet moose-brush and deep swamps, and yet we made five and a half miles; but as the previous disagreeable day on the trail had also been on Friday, it gave us something to be superstitious over.

The packers and Russell got away early next morning, the latter having found a place where he could fell a tree over the stream for a bridge. That morning a couple of ptarmigan came into camp and strutted all around it, evidently aware of the fact that the shot-gun was ahead with the advance party. That forenoon we saw a great many of them. When frightened up, they invariably would light upon some distant snow-drift, for these, too, were plentiful on all sides as we ascended toward the great divide between the Pacific Ocean and Bering Sea drainage basins.

At noon we reached the timber limit, the pass directly ahead and in full view. Through it we soon saw Russell coming back to the divide. He had killed thirteen ptarmigan and three marmots. I determined to call this pass the "Ptarmigan Pass,"[24] from the great number of these birds seen here. Once on the high divide out of the timber, the wind, sweeping away the mosquitoes, felt very refreshing, and had it not been for an approaching thunder-shower I should have tarried there longer, for the camp was in sight about a mile beyond in a little basin. This shower was nearly as long as the one the day before, and even fiercer, the hail actually covering the ground. Again we heard of the great dusky-thunder-bird of the Takous from our Indians, but none of us saw it, it is needless to say.

After it cleared up, the ptarmigan could be heard cackling in all directions, and a half-dozen were in sight at once, through the field-glasses. Their favorite strutting-places seemed to be on the snow-banks near a clump of willows, breaking through an edge of the bank. If they had kept away from the dark-colored brush their almost pure white plumage would have rendered them practically invisible on the snow-banks, however much they may have waddled around over them. I had some doubts at the time (and even have some yet) whether these white grouse were the genuine Arctic grouse (polar ptarmigan) or not. There

---

24. Schwatka's name, "Ptarmigan Pass," has been retained on recent maps.

may be no specific difference, but in several small particulars there seemed to be variations. These grouse have a very coarse, discordant cackle that attracts attention to them at once, and as they are loth to fly when approached, but walk along on the snow or ground a number of yards before flushing, they fall an easy prey to those who will take advantage of still or ground shots. Whether flushed by an intruder or flying of their own will, they always described a curve so uniformly similar that I felt justified in roughly sketching it and dubbing it the "ptarmigan curve." It is similar to that of the grouse family in general, I believe, but as these are so frequently broken by timber and other obstacles, they are not so noticeably uniform. The ptarmigan, on the contrary, is always seen in timberless tracts, although frequently covered with patches of willows and other Alpine or Arctic shrubbery. It is a singular fact that the ptarmigan are most numerous on the Alpine heights, at the headwaters of the great Yukon River, and on the low, level plains at its mouth, with very few of the birds in between. The Esquimaux at and near the mouth of the Yukon snare these white grouse by the sled-load, selling them at five cents each at St. Michael's and other near places. They are snared by an ingenious arrangement of cut-brush fences directing the birds into the traps.

The doctor made a few hours' trip to the top of the nearest high hill, and reported, upon his return, that from this point no less than fifty-five lakes could be seen in the great valley beyond, one of which was undoubtedly the big lake for which we were striving. He killed two ptarmigan and saw a number of woodchucks in a rocky place on the hillside. The crops of all these ptarmigan were filled with willow-buds mostly, and with a few hardy Alpine berries.

There was an enormous drift or bank of snow at the foot of the hill near camp, and it was blotched with red, looking as if blood had been spilled in places and then had about half faded out. It was evidently the red snow, a sort of algae, so frequently seen in the Arctic and on high mountain slopes. This camp was right in the midst of a marmot "patch," to use a miner's phrase, and these little creatures could be heard chirping at all hours until we got away. The Indians are very fond of them, but our surplus supply of ptarmigan did not tempt us to feast on any of these rodents. These natives eat a marmot by roasting it whole, and, when swollen up almost to bursting, they cut a small hole in the belly, remove the entrails, and then close the hole up with a wooden skewer, eating the entrails while the body is cooking on a spitted stick. These marmots are sufficiently numerous throughout the Alaskan mountains to prevent starvation of one or two good hunters going through them on an ordinary journey. They are ordinarily snared by the Indians, since, like prairie-dogs, they generally keep so close to their burrows that, unless instantly killed, they make good their escape.

Seeing the big lake so inspired the Indians that they were up bright and early next morning, and got away with the first packs before three o'clock. I think they intended to make the lake that day, but there were two more camps ahead to fight the mosquitoes in before we made it. There is no distance so deceptive as that estimated from a height looking into a valley. It sometimes seems a jaunt of an hour or two to a point that will really require a hard day's tramp to reach.

I noticed that morning, as we started, that many of the lakes on the high divide had still a fair share of their winter's covering of ice on them, in some places one or two feet thick, but apparently rotten, while many of the streams were crossed by taking advantage of the snow-bridges over them not yet tumbled in. Ptarmigan were on all sides, right and left, before and behind and overhead.

About nine, there burst on our view the whole valley of this great tributary of the mighty Yukon River, the scene ahead being perfectly studded with lakes that shone like diamonds, pearls, opals and moon-stones in a field of emerald, as the morning sun's light was reflected from their surfaces at a dozen different angles. It was truly a land of lakes, with a full half hundred in sight. A part of the big lake—probably a third, as it proved afterward—was in sight, strung out like a great glistening snake, and at the near end the little bight where the land-labors ceased, and we took to the water again as far as old Fort Selkirk on the Yukon River. The lake seemed almost under our feet, not over three miles away, and I believe that had the Indians thought it yet three days away they would have succumbed even at that late hour. They were certainly very sick of their bargain, nine-tenths of the cause lying in the double packs, for I think the whole trail would be an easy one for Indians accustomed to this work, as the Chilkats and interior Takous, with but one load to carry. Doubling a trail is always discouraging, and I think the more so as we descend in the degree of intelligence, where the reasons for not doing it are not so apparent to the laborers. A portage of two or three miles is bad enough, and delays an expedition out of all proportion to its length, but one of eighty to one hundred is almost a good summer's work in itself. The old Canadian *voyageurs* so dreaded the time lost at them, not the labor, that their method of procedure was to put two hundred pounds on the back of the packer, and *run* him across the portage both going and returning, for all distances less than about five miles.

That day the Indians were without lunch, and were correspondingly cross and irritable. It was caused by Russell giving them a camp-kettle of beans in the morning, to be saved for that meal; but happening about breakfast, the beans disappeared along with it, Indian-like.

Some days back, I had sprained my ankle—I think it was the time I had on my extra slippery shoes—and though it hurt me considerably at the time, I paid but little attention to it for the next few days. It was now paying me considerable attention, however, and growing steadily worse; so that by the time the lake was reached I was just about *hors de combat* in a pedestrian way. It was one of the luckiest unlucky accidents I ever had happen in the wilds of a new country.

That evening we went into camp in a deadwood patch, after clearing eight miles over a bad road. Indians going ahead reported the first little lake but a mile distant. That day I killed six white and one dusky grouse, while others secured some also.

Despite the dense swarms of mosquitoes that night, the Indians were up and away with first packs by 3:30, the first lake being reached an hour later.

As the folding canvas boats now enter our story more closely, a very brief description of them would not be out of place. They were made by Mr. King, of Kalamazoo, Michigan, and in their construction are quite unique, having a strong wooden keel with "lap-streak" side braces, ribs and gunwales of tempered steel wire. With all this is coupled a double jacket, both of which had to be ripped to sink the boat. The inner one was perfectly water-proof, and if water got between the two, as it usually did, it made no difference beyond the additional load thus carried. In short, they were very strong boats, capable of carrying a ton each, and weighing a trifle over one hundred pounds apiece.

I had named them:

Andrew Allen Bonner,
Robert Edwin Bonner,
Frederick Bonner,

after the patrons of the expedition—the proprietors and editors of the New York *Ledger*.[25]

At "First Lake" we put up the *Frederic Bonner* in forty minutes (I think thirty were used in fighting mosquitoes), got in all the effects, about one thousand pounds, and with the doctor and I as navigators, got away at 6:20, the Indians and Russell making overland. "First-Lake" was two and a half miles long, and we paddled it in forty minutes, doing some unrewarded fishing on the way.

The first interlacustrine portage was about half a mile over, and "Second Lake" gave us only one half a mile. The second portage was a mile and a half long, while "Third Lake" gave us two and a half miles, a good breeze to sail it with and keep off the mosquitoes, and a two-pound pickerel,[26] another of about the same size getting away. Here also we met our last party of outward-bound fur-traders, our packers securing some moccasins and a little dried moose-meat from them. They were wild enough creatures to suit any one, the grown ones extremely shy, while the dogs skulked away growling, and the children ran howling from us as soon as the parties met.

On the third portage (two and one-half miles across) I came to a pretty specimen of protective mimicry in spiders, one of these small arachnids on a dwarf sunflower being exactly the color of the blossom on which it was hidden; the first pure yellow spider I have ever seen. The varieties of mimicry assumed by this family are known to be wonderful. Scientists are constantly finding new cases, but I knew of none similar to the one found here. The following, from *Science* of July 7th, 1891, just a month later than the above-mentioned time, is given as recent interesting examples:

> In the journal of the Elisha Mitchell society, Mr. Atkinson calls attention to two new cases of protective mimicry in spiders. A *Cyrtarachne* takes shelter in summer and autumn under leaves, where it has absolutely the aspect of a small univalve mollusk, which is extremely abundant, and which often fixes itself in an analogous position. The

---

25. Schwatka names three folding boats but apparently only two were used. (See Hayes 1892, p. 121). The boats were designed by Charles King and first manufactured in Kalamazoo, Michigan in 1889. They were used extensively during the Klondike rush of 1897–1898. Improved versions continued to be manufactured in Kalamazoo until the 1970s (*The Kalamazoo Gazette*, Jan. 26, 1991; Feb. 17, 1985).

26. The "pickerel" caught by Schwatka was probably a northern pike (*Esox lucius* L.). (Scott & Crossman 1973 pp. 346–375.)

second example is found in a small spider, *Thomisus aleatorius*, which is remarkable for the length of its forelegs, the hind ones being on the contrary, very short. This spider, which lives upon grasses, ascends the culm, stops suddenly, and disappears from sight. It suffices to fasten itself to a spike by its hind legs, and to bring together its forelegs, extended, and form an angle with the culm in such a way as to make itself nearly undistinguishable from the spikelets.

The "Fourth Lake" gave us an hour and a half on its waters, and as this brought us up to seven o'clock we went into camp. Although the packers had carried only at the short portages, their ankles were black, bruised and swollen out of all shape, and it was evident that we were not reaching the end of their journey any too soon. Their foot-gear was nothing now but a mass of tied-up rags. Russell said that the mosquitoes were the worst he had yet seen on the last overland trip around the "Fourth Lake." With both hands full of willow-brush applied furiously he could not keep them out of his face.

The fourth portage was but a mile, and "Fifth Lake" the same distance, while the fifth portage was only a third as far. This over, and by the middle of the forenoon, we stood overlooking Ah'k-Klain, which, translated, means "The Big Lake" in the T'linkit tongue; (*Ah'k*–Lake; *Klain* [Klane]–Big).

I had thought the packers would take a rest for a day or two, or at least for the afternoon of the 16th, but they were so anxious to overtake the last party of Takous we had passed that they got away at three that afternoon. They had performed a herculean task indeed, and deserved well of the fruits of the undertaking, whatever the future may show for this.

And now, on the shores of the great inland lake, it is well to take a retrospect of the trail, now for the first time mapped, and see whether there is anything in it either for the present or the future.[27] The trail was eighty miles long and some forty-five on the river, or a rough one hundred and twenty-five miles altogether. I believe that a powerful light-draught river steamer can ascend to the head of canoe navigation, or at least to the confluence of the main branches; still, the report of a good

---

27. Schwatka sent a letter describing the Taku Trail to his backers in Juneau with the returning packers. It was published on the front page of the *Juneau City Mining Record* nine days later (June 25, 1891). Hayes' map of the route (Yukon District—Sheet 1), was published in *The National Geographic Magazine* the following year (Hayes 1892, plate 19).

river-boatman for all navigable months of the year would be the best information on which to test this, and would not be hard to obtain. From either point a pack-mule trail to Ah'k-Klain can be easily made. With pioneering parties of fifteen or twenty men, I have seen pack-trains taken through equivalent country, keeping up with cavalry commands averaging over twenty miles a day. Such a pioneer party ought to make a very good trail here in two weeks, or a month at least. From the big lake there is uninterrupted river navigation to Bering Sea in the summer months; although it is possible that one point—the Rink Rapids—may give trouble at certain unfavorable stages of the water. It is, therefore, a mere matter of sufficient commercial interests being developed in the Yukon basin to justify the making and maintaining of such a trail, in the spring and summer months. The city to be benefited by this in Alaska is plainly Juneau.

But the greatest benefit this route will ever confer is, I think, in a railway, and not so very far in the future either. To the casual observer, limiting the considerations to the country directly drained, there is not a great deal to tempt a railroad into Alaska; but as two-thirds of our Western States and Territories were developed by railways, looking not to what was in them altogether, but to some great terminus beyond, so Alaska may hope for one on the same principle. The great Siberian railway terminating at Vladivostock is assured, and will be built, it is officially asserted, in twelve years. The nearest perpetually open winter harbor to it in America attainable by a railway, is in the Juneau vicinity, and by a railway down the Takou River. There is absolutely no point west of this, unless Bering Strait, thirty-six miles across, should be tunneled, a far too formidable undertaking for the advantage to be gained. From Juneau to Vladivostock the sea voyage can be made by an Atlantic "ocean greyhound" in a week, and at all seasons of the year. The construction of a railway along the Takou, through the Alaskan coast range, would cost less, I believe, than the average road through the Allegheny Mountains. The great inland plateau gained, the construction to any point on the Canadian Pacific would be about equivalent to that on our great Western plains, of which this plateau, after all, is but a northern extension. The objective point on the Canadian Pacific to strike would most likely be Winnipeg, a busy, solid city, with many direct railroad connections to the very heart of the United States. In fact, this Alaskan road would probably be a natural branch of the Canadian

Pacific, that could operate it to better advantage than could be done as an independent line. This branch would be less obstructed by snow than the main line, which, again, has less nival difficulties than the Northern Pacific, the line of maximum snow obstruction for North America lying somewhere between the latter road and the Union Pacific still further south.

SIXTH LETTER

I do not know if the title of Ah'k-Klain ("The Big Lake" in T'linkit) amounts to the full signification of a proper name or not, or whether it is merely descriptive. I know of no other title given it by these or any other natives; while, on the other side of the question, there are several Ah'k-Klains (or big lakes) in this general locality, which would hardly be the case except as a general descriptive appellation. Had we taken the northern path at "The Trail Splitter"—which we came near doing by mistake—its course would have led us to another Ah'k-Klain, which is simply the largest of a series of lakes on that trail, according to the Indian version.[28] Again: Lake Bennett (which I named after James Gordon Bennett in 1883) is Ah'k-Klain to the Chilkats, a band of T'linkits that trade in the interior by a trail that takes them over this lake and several others, of which Bennett is among if not the largest. So several more or less clearly defined Ah'k-Klains have been known as existing hereabout for varying periods, and some of them have crept into those usually erroneous maps constructed from native information, but nearly always as *Lake* Ahklain (with several methods of spelling it), a tautological error about equivalent to *Lake* Lac Qui Parle or Rio Grande *River*. Outside of a guttural explosiveness on the first syllable *Ah'k*, it is a very pretty name, however; much prettier, in fact, than T'linkit names in general.

One of the attractive features of its shores was the large number of wild roses seen in bloom, and that crowded all the open spaces where

28. The "Big Lake" (Ah'k-Klain) reached by Schwatka is now known as Teslin Lake. The native name, Ah'k-Klain (now reduced to "Atlin") has been applied to a second "big lake" west of Teslin Lake. The third "big lake," located westward of Atlin Lake, once known as "Lake Takou," is now known as Tagish Lake, its largest arm known as Taku Arm.

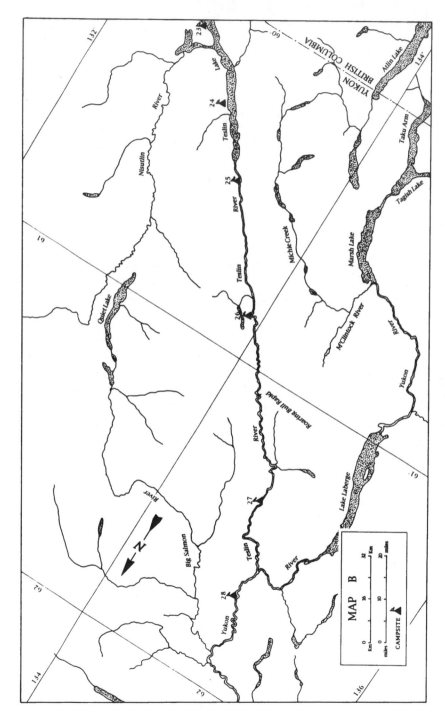

Map B: Route of the New York Ledger Expedition from Camp 23 (June 19) to Camp 28 (June 24). Topography based on Canadian Geological Survey Map Sheets: Atlin (104N), Skagway (104M), Teslin (105C), Whitehorse (105D), Lake Laberge (105E), Quiet Lake (105F), Glen Lyon (105L).

timber-fires had killed the trees and allowed the sun to get freely at the soil.[29] The robins and the roses were a graceful contrast to the snow-banks and ptarmigan we had so recently left behind. But there is no rose without a thorn, and there were certainly a thousand thorns in this case for each rose, if the mosquitoes can be figuratively spoken of as such. Russell and Hayes tried to take a short hunting and fishing tour into a little bight of the lake, and were driven back by these numerous pests.

Most of the 17th of June was spent in rigging out the folding canvas boats for lake navigation. Masts and stays were made from poles and boards, and fish-slicker blankets were extemporized into sails. Two of the boats were thus fixed, the *Robert E.* and *Frederic Bonner.* Very good oars were made by Russell from rough slabs,[30] he being a carpenter and boat-builder. A fine breeze all day from the south raised our hopes into believing that it was somewhat permanent, for the lake from here trended to the north-northwest. It was a long, narrow lake, seldom over six or seven miles in width, stretched out along the course of the river—a sort of a riparian lake, as some physical geographers would say.

We had not struck the upper end of it, as some may imagine from my description, but on the western shore some distance from that point. To determine how far it was to this end and make its survey complete, the doctor and I started for it in a boat the afternoon of the 17th; but after paddling vigorously half the afternoon against a head wind we were forced to turn back, the lake still stretching out southward, around a slight bend, as far as the eye could reach.

This camp was the farthest point toward the east that we reached on the trip, and I was correspondingly anxious to get an observation for

---

29. The rose mentioned here was probably the Prickly rose (*Rosa acicularis* Lindl.). (Viereck & Little 1986, pp. 187–189).

30. The slabs used by Mark Russell to make oars may have been taken from an abandoned saw pit used the previous year by his prospecting party (Hayes 1892, p. 119). Other members of this party included Messrs. Lowe, Lenox, Brown, Konnel, and Houser (*Juneau City Mining Record* Aug. 7, 1890). Over the years many prospectors built boats at the head of Teslin Lake from whip-sawn lumber. For example, Mike Powers and associates built three boats there in 1876 or 1877 (Dawson (1987, p. 156B). Powers later staked claims near Juneau and was associated with the legendary Lost Rocker Mine. He was killed in a cave-in on one of his properties in Silver Bow Basin near Juneau sometime during the winter of 1885–1886 (DeArmond 1967, pp. 187–188).

longitude; but while successful as to latitude, the weather was not favorable for an equivalently good one for the other and more important co-ordinate.

It was quite evident that the lake, like the rivers we had met on the Pacific Slope, was far above its normal level. All of the shore-line timber was half under water, showing that the lake-beach was well submerged, and it was only at the very open places that we could launch or land our boats favorably.

During the day we saw a heavy signal smoke far down the lake, showing that all the Indians had not left the country, despite the large numbers of outward-bound fur-traders we had met on only one trail. In fact, I was greatly surprised to find such a large number of T'linkits (these were all the Takou tribe of T'linkits) making their permanent homes on the British American inland plateau. I knew well enough by personal experience that several of the tribes had trails leading thereto, but I had supposed that they only used them to make fur-trading excursions to the inland tribes, and in no cases to gain their homes. I believe yet that the Takou T'linkits are probably the only ones not confined to the Alaskan coast or the rivers tributary thereto. The Takous have a legend, so my own Indians told me, that all the T'linkits were once Takous, and lived like friends and brothers in one big village on the banks of the Takou River, until a quarrel arose (it may be a waste of printer's ink to state there was a woman in the case), and many were killed and maimed therein. As a consequence, a number of discontented parties radiated out from this T'linkit Babel and founded the various clans, or sub-tribes, of Yakutats, Chilkats, Hoonyahs, Sitkas, Kootznahoos, Auks, Stikeens, Keh'ks, Kous, Tongass and others of the T'linkit tongue.

Beyond this country lies that of the Taltans, and the single Indian of that tribe among my packers, Barney, told me that it was a fair land to look upon, much finer in nearly all respects than the one we were now traversing. It was stocked with moose and caribou that waded through sweet grasses knee deep, while such a thing as a famine, or even noticeable want, was unknown. How much of this was patriotic boasting and how much the truth was hard to tell; yet, from his general character and that of the land adjacent to that he described, I think most he said was true, or at the worst, only a little high coloring of the truth. From Ah'K-Klain a trail, wagon-road or even railway could be run practically in any direction that bore the cardinal point of *eastward* in its course. From here it is but a comparatively short distance to the Cassiar placer mines of British Columbia—mines that the most experienced declare are not at all worked out, except by the exorbitant freight-rates to them, which bars everything unless phenomenally rich. A railway through this country would open a large number of such mineral showings.

Whenever two clans or branches of the same tribe of Indians are found living inland and coastwise respectively, it is nearly always the case that the interior portion will be the better physically and mentally; and this, I believe, is true of the two Takou divisions as far as I could learn and see on such a short acquaintance. The reasons are obvious: The inland native is a hunter of large game, and his muscles and mind are developed by the chase; while the coaster is a fisherman, requiring comparatively little exertion, and even his food is sometimes placed directly before him, as the shell-fish along the shore. "When the tide is out the table is spread," is a saying that explains it pretty fairly. Where there are wholly different tribes, as in the case of the coast-abiding Chilkats compared with the interior Tahk-heesh, this condition may be reversed; but where all other things are equal, I have never yet seen it fail.

About the middle of the forenoon of June 18th we got away under "slicker" sail with a fine four-mile breeze. It was luxury, indeed, to lie stretched out on our bedding and knock off, in an hour or two, the equivalent of a day's march on the trail; while as for a day on the lake, it nearly equaled a week on the land, and there are no figures in mathematics to express the comparison as to labor. In an hour and a half we were abreast of Sour Dock point, six miles away, this title being the one Russell gave it as a remembrance of the year before, when this acid-tasting weed, in the shape of pots of greens, furnished his prospecting party with several palatable dishes to mix with the everlasting pork and beans of such an expedition.

Just beyond here there is a tumble-down Indian "shack" on the east bank that may be inhabited occasionally, for I had found out by this time, that however dilapidated a native building may have looked, especially about the roof, it was no positive sign but that that part is repaired annually, and at certain seasons it is occupied. A heavy, roaring sound from this bank indicated that a large, swift stream came in near by, but its mouth could not be plainly made out. Russell said its mouth was about fifty yards in width, its bed shallow but rapid, while its delta was clogged with drift-timber that it had brought down.[31]

Everywhere along the shores of the lake this drift-timber of all dimensions was to be seen. The winds of the locality seemed to have more influence in directing its course than the very slight current produced by the draining river at its northern end; so that while large amounts of it were thrown on to the lake by the incoming rivers, but little of it found its way out. Some of this must have been drifting around for years, bumping against the banks, so denuded were the trunks of bark and fairly glistening with the polishing they had received. Whenever we tried to land for lunch or camp, this driftwood on the lake, at this high stage of water, barred our way more or less effectually, and we were lucky, indeed, if a place could be found where the driftwood was solid and we could carry the boats ashore from its outer margin. In nine cases out of ten, however, the outer fringe of logs were loose in the water, would roll whenever stepped upon, and this, combined with their slippery, slimy sides and our stiffened condition, just crawling from the boats after several hours' confinement, made our landings a labor and a difficult undertaking.

Just before noon our wind died out, and the rest of the day we spent at the oars. At noon we passed another rapid stream, coming in from the east, near a prominent broken bluff, its mouth apparently fifty to sixty feet in width.[32] The middle of the afternoon a wind set in from the north, which in a little while became quite stiff and delayed us considerably. This alternation of forenoon and afternoon winds in opposite directions we found to be quite common, but with considerable diversity as to the time of day when the change took place.

About four o'clock we reached Cliff Point on the east side, the first abrupt, broken promontory on the lake that we had met. It was of granite—so the doctor's investigation showed—about seventy-five feet perpendicular, and from nineteen to twenty miles from the morning camp. It was a most picturesque break to the gentle gradients so characteristic of the shores of Ah'k-Klain, and a very conspicuous landmark on the length of the lake. The low, flat hills bordering the west shore as seen from here, were most beautifully variegated in green, black and

---

31. Now shown on modern maps as Jennings River.
32. Now shown on modern maps as Swift River.

red, showing the contrasting colors of the living timber, the burnt trunks, and where the leaves were only scorched.

We went into camp quite early, but with a great, big twenty-three miles to our credit, a distance that appears insignificant now, looking backward from the land of railroads and ocean greyhounds, but that made camp seem like a foreign country from the place we had left in the morning by comparison with our former gait. In addition to a wide fringe of floating timber at the shore, we had to contend with a broad strip of half-submerged, thick brush before the evening camp could be made.

The 19th was almost a repetition of the day just described, except that the wind changed much earlier, and we consequently had more rowing. The first group of islands, quite prettily nestled in a little bay on the west side, was passed early that morning.

About noon, looking backward as we sat at the oars, we could see a natural water horizon of a few degrees, the lake looking as if it opened out to sea.

The demarcation line between the burnt and living timber on many slopes was so straight and well-defined in long east and west courses that I thought it could only be accounted for by the changes of the north and south wind, apparently so constant here.

About noon that day we came abreast of the first peak of the Caribou range. This is a most conspicuous, isolated cluster of peaks[33] on the west shore, around which the lake slightly bends, as if divided at this point into two great arms. This is the name given it by the Indians, and although it is one that occurs with great frequency in the northwestern part of North America, due to the abundance of the noble game animal it represents, yet this is so far from any named districts that it is probably better to retain it.

The first day's boating had not given us a single "rise," with three trolling lines set out; but this day, when we saw some fish jumping, the trolls were again cast, and two salmon trout, one of five to six pounds and the other nearly twice as heavy, rewarded the fishermen.

---

33. Schwatka's "Caribou Peaks (Range)" are shown on recent maps as Dawson Peaks, and are known locally as the "Three Aces." During construction of the Alaska highway they were known briefly as the "Five Aces." According to Native legend they served as a refuge for animals and people during a great flood long ago (Teslin Women's Institute 1972, p. 21).

Just about two in the afternoon a furious squall set in from the west, that soon raised heavy seas and white-capped waves that tested the folding canvas boats quite thoroughly, and much to our satisfaction. We pulled for the east shore, and here, back of a half-submerged point of brush and tied by long painters to a cottonwood tree that seemed to grow out of the very lake, we rode out the storm that, a half-hour afterward, died down as suddenly as it had come up. It was due to a thunderstorm that had passed far west of us, but we always took warning at their approach thereafter.

Again, about four, we passed high, steep cliffs of gneiss and granite on east side, and, just beyond, a river coming in from the same direction, with a mouth of about one hundred yards in width, and emptying into a large bay that put back from the main lake.

With a 27-mile record that day we felt like camping early, but the driftwood vetoed our efforts, and warned us convincingly that if we wanted to get all work done before dark we must hereafter seek an early camp to compensate for time lost in this way.

After passing the Caribou Peaks our course bent slightly more to the westward. We got away early the 20th, as a spanking breeze our way induced us to press matters while it lasted.

About the middle of the forenoon we passed the mouth of the Heen-Klain (Heen-River, in T'linkit) or Big River of the Indians.[34] It comes in from the east, and is about one hundred to one hundred and twenty-five yards wide at its mouth. Its valley is very conspicuous, and can be readily traced back inland for forty to fifty miles, large snow-clad hills flanking it on the southward. It is evidently the largest river draining into the Ah'k-Klain, as we were careful to ascertain from reliable Indian accounts. On these long, narrow riparian lakes, or those that are simply expansions of a river's course and generally caused by a damming at the lower end, one usually expects, and nearly always finds, the main incoming river at one end and the draining stream at the other; but this is an exception to the rule. If the river at Ah'k-Klain's head is of any considerable size, which is likely, then the lake receives quite a number

---

34. This is now known as Nisutlin Bay. The Nisutlin River enters the head of the bay about six miles to the east. The village of Teslin, Yukon Territory is located nearby. Teslin was founded in 1903 as a trading post by Tom Smith, a resident of Dawson (Coutts 1980, p. 263). The Alaska Highway bridges the bay one mile east of Teslin (km. 1292).

of fair-sized streams to pour into its draining river, and can hardly be called a true riparian, or river lake.

While off Heen-Klain's mouth the wind died suddenly down, being somewhat earlier than usual. Here a couple of Takou Indians, in a light, birch-bark canoe, paddled over to us from the west side, having previously signaled us by a gun-shot. They had some dried moose-meat with them, and though it was a long way from inviting, we bought the best-looking piece among the lot. Our conversation was rather limited, but we managed to understand enough to make this simple barter; and they were bright enough to get the better end of it.

About an hour before noon we passed a low point that did not seem to be fortified with a wide *cheval-de-frise* of drifting timber, so we got ashore here to obtain an observation for latitude and lunch on some of the debatable moose-meat.

That evening no camp could be made until we had removed about twenty feet of crosswise floating timber to get to it. Then it was an exceedingly poor camp when we did get at it, and in strange contrast with the other fine ones we had on the lake shore. That evening's meal had the best the market afforded, there being gulls' eggs, moose-meat and salmon-trout, with pork and beans for dessert. The record for the day was twenty-five miles, the greater part being done by good, honest rowing.

The next morning the drift was as bad as ever, and the lake apparently higher. In fact, we rowed over a number of submerged flats that forenoon, the trees and brush half in the water showing its high condition. This Sunday gave us the first steady all-day breeze we had had, and it was dead ahead.

There is one peculiar condition of the mountain-tops in this general locality that I had noticed for the last ten days to two weeks; and in this I refer to the large number of notched peaks to be seen, and especially abundant as seen from this lake, three being in sight at once. I know that as hills or mountains are domed, rounded, broken, etc., geologists can make fair inferences as to their composition, but I know of nothing shown wherein every third or fourth sharp peak among a large number looks like a huge hind-sight to a rifle.

By two in the afternoon the lake began to narrow rapidly, and the outlet could be seen among a lot of white, broken banks, showing the effect of a contracted, swift current. At 4:20 we entered the draining river, and once more took up a good gait, with the consolation that even a head wind, unless amounting to a gale, could not wholly counteract. We went into camp at once, but it was so very poor, being half under water and the other half under mosquitoes, that, after supper, we pulled out and floated down three miles to a much better locality.[35]

That day we made twenty miles on the lake, or ninety-five in all from our first camp on its western shores. We proved that it extended at least six to ten miles farther southward than this latter point; so one can be safe in asserting that Ah'k-Klain is over one hundred miles in length, and will, therefore, take its place among the great lakes of the British American Northwest—a land famous for its large lakes and diverse lacustrine systems. I know of no lake to the northwest of it that is larger on our continent, so it may be considered the grandest outlying body of water as approached from the Alaskan side, although not of such comparative importance alongside of the Great Slave Lake, Athabasca, and others lying to the eastward.

Several times I came in contact with the fact that these inland Takous knew more or less of another T'linkit band, called the Stickeens, and showing that a probable interior or inland communication existed between them. These Stickeens are in greatest force around and near Wrangell, Alaska, the importance of this town depending on the Cassiar mines of British Columbia, of which I have spoken, and reached from Wrangell by the Stickeen River. This stream (strictly the Stückheen— German ü—in T'linkit meaning something equivalent to Swift Water) is navigable for one hundred and forty miles, and requires four or five days for its ascent by a river steamer and but one for returning. Then there are some seventy-five miles by pack-train to Dease Lake, and twenty-five miles over it to Cassiar. All through this land the Stickeens have been employed, and many have extended their trading excursions to the Takou land of lakes.

This Stickeen, or Stükheen, is one of the most picturesque rivers in the world for its length, and it seems that it ought to pay to put an excursion steamer on its waters. The P.C.S.S. Co. sold 5,021 tickets to Alaska in 1891, and enough ought to be obtained from this large number to pay for such a delightful side-trip. About forty miles from this river's mouth there can be seen the curious sight of a boiling spring on

35. Schwatka's first camp on Teslin River (Camp 20) was located near the present Alaska Highway bridge at Johnson's Crossing, Yukon Territory (km. 1346).

one side and an enormous glacier on the other, the latter attraction being but one of a score or so of gigantic streams of ice along its course. The Stickeen River freezes every winter to its mouth, and can then be ascended by sledges, opening about March 1st. The Indians have a legend that the great glacier of the Stickeen once put across the river, which flowed underneath it, and through this icy tunnel they once sent an old man in his canoe. To corroborate the first theory, there can still be seen the remnant of a glacier opposite the main mass of ice; but there is nothing to corroborate the second part, except the ample inherent meanness of the people to do such a contemptible trick, especially if the old man was a decrepit and valueless slave. Some time along about the latter seventies, Shotrich, the great chief of the Chilkats, of whom I have already spoken, killed sixty slaves before Shakes, chief of the Stickeens, who, not to be outdone, slew sixty-five before Shotrich. Then the two clans fought, and the Stickeens were whipped, my informant adding that Shakes committed suicide, as a result of the combined loss of the slaves and the slaughtered.

Kat-o-shan, the present Stickeen chief, gave me some very interesting information as to the use to which the T'linkit totem-poles are put, and regarding which there are a number of theories. Calling his attention to the back of one in his village apparently hollowed out and the excavation boarded up, he said they are often used as burial-places for the ashes of cremated Indians, while others are geneological and historical. An Indian erecting a totem-pole gave a great *pot-latch* (the Chinook jargon for *giving*) in doing so, and got the members of other clans to do the work, for fear of being ridiculed, until the ceremony was ready to begin.

There is fine hunting around the head of navigation on the Stickeen and in the Cassiar region, there being caribou, moose, bears, mountain goats and sheep.

SEVENTH LETTER

The last letter left us at the entrance to the draining river of Ah'k-Klain. This stream has a number of names. When I passed its mouth in 1883, as the first person in the country bent on explorations, I called it the Newberry River, after Professor Newberry, President of the New York Academy of Sciences; the miners of the region, the only civilized people in the land, and the first to traverse its length, call it the Hotalingua; while the Indians have different titles according to the tribes. The Tak-heesh living at its mouth had such an unpronounceable name that I felt justified in changing it. The T'linkits call it the Tes'l-heen (on some maps as Teslin) the equivalent of "Rapid River;" but believing that they were only travellers and traders in this country, my first inference, that of 1883, that their names were not to be given precedence over local Indian names, will have to be abridged somewhat, since they, at least, live on its outlet.[36]

At our first camp on the Newberry it is about one hundred and twenty-five yards in width, and running from four to five miles an hour, but probably a mile or so of this gait was due to the then high stage of water, and the normal or average current is that much less.

There were thousands of swallows making their homes in the white clay cliffs of the river-banks, while they could be seen flying in every direction. I noticed recently the comments of a traveller in a northern region who asserts that the dense swarms of mosquitoes so common in the neo-Arctic and sub-Arctic regions are always found in this abundance just beyond the northern limit of the swallow's homes, and from which laudable inferences are drawn as to the usefulness of these swift

---

36. Despite Schwatka's insistence on the name "Newberry," the name never caught on and Teslin remains the river's official name (Coutts 1980, p. 264).

Two inland Tlingit Natives in bark canoe, Teslin Lake near Nisutlin
Inlet, June 20, 1891. (U.S. Geological Survey photograph by C. W.
Hayes) [Hayes #316]

and graceful birds as mosquito destroyers. Now I found them here living in comparative harmony, or, at least, the large number of swallows
made no appreciable inroads on the pests.

Compacted sand-banks were numerous along the river, and they
were stiff enough in many places to present almost perpendicular faces
from twenty to fifty feet in height; in fact, almost amounting to a friable
sand-stone. All along these faces the loosened sand was pouring down
in streams of variable size, from that of the finger to sand-slides that
sent up huge clouds of dust. Well-marked terraces were to be seen
during the forenoon. Along the foot of the sand-slides there could often
be seen a white efflorescence very similar to the alkali deposits so common on some of our Western plains, and evidently of the same general
nature.

It would be tedious to mention all the tributaries we noted that day,
but their combined effect was apparent, for oftentimes the river broad-
ened to about two hundred and two hundred and fifty yards before the
evening camp was made.

Many ducks and geese were on the river, but it was now so broad that we could not get near them and keep the swiftest central current, for they certainly seemed the wildest, wariest birds I ever saw, to be hunted as little as they are in this out-of-the-way corner of the earth. Several shots were taken, but few of them being effectual, so dense is the plumage of these northern aquatic fowl.

Late in the afternoon we saw two large land or sweet-water otters among the submerged brush of the west bank. They seemed to be playing in the swamp-like thicket as we rounded a point of timber projecting into the water, and both parties were so surprised at the sudden encounter that each stared the other out of countenance before the brutes realized that there was any danger, or the humans that they had excellent weapons to make it dangerous with. We did not bring any otter-skins home with us. I have seen a number of sweet-water otters trapped in the inland waters of the United States, but I had never seen such huge beasts as these before. They seemed more like aquatic panthers playing in the waters than the animals I had been used to.

There had been a number of showers during the day, and so threatening at night that, for the second time on the trip, the tent was pitched; a record of thirty-one miles behind us.

Early in the forenoon of next day the sand-bluffs along the river changed to gravel cliffs, but quite as steep and precipitous as before. We could also notice that the waters of the Newberry were getting perceptibly muddier from the effects of washing the banks and the receiving of tributaries having no lakes upon their courses. However muddy the waters of a stream may be, even the fine, impalpable, chalk-like silt in a glacial eroding river, a lake of any comparative size upon its course is a perfect filter, figuratively speaking, and the draining stream is as clear as crystal, even if the incoming waters be colored with sediment.

The first indication of Indians on the river was not one of life, but the reverse. The middle of the forenoon we passed a grave on the east bank. Over it was a wedge or "A" tent, with two shaved logs, ten to twelve feet high, alongside. It showed recent work about it, and, I think, was a Tahk-heesh or "Stick" interment, rather than a T'linkit so far down upon the river. I had heard considerable before this, as well as afterward, that the epidemic of influenza, or Russian influenza, so commonly called the "grippe," had devastated a number of Indian tribes in Alaska and the British American northwest, both on the coast and inland; so this recent burial may have been a result of its travels. There are very few Indians in this country, probably one to a square township, so that even the ravages of an epidemic would not show much sign to a traveller pursuing a single course through the country. On the lower Yukon River, where the natives are much more numerous, I heard, in 1883, of a great epidemic that had swept the country, but only found a half-dozen or so buried in one place to show for this particular incursion of the disease, and some of these, I think, were older graves.

Near this "Stick" grave the bluffs were very high and picturesque on both sides of the river, and looking as if the Newberry was cutting through a range of hills. Islands were getting numerous, and an effect I noticed in 1883 was showing in regard to them. I refer to the enormous amount of driftwood piled on their up-stream ends, while there is none on the lower parts; so if a traveller on the river looks down the stream he sees a lot of desolate driftwood everywhere, above which the green timber appears, while casting the eye in the opposite direction he sees a most picturesque grouping of islands that are pleasing by contrast.

About noon I came on an old familiar friend of my former Alaskan travels. I refer to a conspicuous white stripe of volcanic ash just between the soil and its substratum of clay, and occasionally cutting through the former. This varied between a few inches in width to as many feet, the thin parts being on the crests of the rolling bluffs, as would be expected, while the thicker deposits were in the lower curves of the banks, as exposed by the cutting of the river. This ash is evidently the product of an ancient volcanic eruption not far away, and thrown upon the earth over a wide area and before the soil was formed. My previous observations were wholly on the Yukon before, and this extends the limits of this ash area much farther to the southward. The original deposit may have been a number of times as thick as the present one, and compacted by the soil forming over it, but it is so loose and friable that it is yet light as pumice-stone, and certainly could not have been compressed an unusual ratio.

Just below on the river a trail crosses that is quite noticeable where it descends a steep cut in the high bank, and without which sign we would probably have passed it unobserved. Small poles stuck at the water's edge probably indicated the crossing, and had we searched the brush near by, no doubt a birch-bark canoe or cottonwood "dug-out" would have been found stored away, to be used as a ferry by the Indians when travelling it.

By far the greater part of the native travelling is along the rivers, and cross-cut trails are by comparison rare. Their principal subsistence is from the rivers and their immediate valleys, while their abiding places are along the streams, so they would rather descend a water-course for a hundred miles and ascend another for a like distance to reach a certain point than to cut across overland for fifty. I do not think it is because they are so averse to walking as it is the carrying of heavy loads on their backs. Their dogs are mostly used for winter sledging, and can seldom carry over forty pounds on their backs, while thirty as an average would be a better estimate; so in transporting their effects in the summer the burden of it falls to the humans unless the rivers are closely kept. Even in the winter the frozen surface of the rivers is the favorite channel of communications, except where a great distance can be saved across land or the way is smoother than the average land travel.

So far along the river we had tied the two canvas folding-boats together for company's sake, doing only enough paddling to point the heads down steam and keep them in the swiftest current. We soon stopped this, however, for in rounding a rough, rocky point we came somewhat suddenly into a short chute of rapids that played havoc with our two boats tied together, shipping water into one and dashing heavy spray all over a lot of books that were laid out for map-making, observations, etc. The boats had stood twice this height of waves before on the lake, riding them like ducks; but tied together, they were as helpless and unmanageable as a raft. We pulled ashore, built a fire and dried out the worst splashed books and papers. These side rapids were quite common during this high water, and were probably rough enough to swamp a badly managed small boat; but there was nothing bad in them to a careful person, while it was easy to avoid them altogether by a little rowing. The worst danger was not in the rolling waves they formed so much as in the high banks they often cut under, and which occasionally came tumbling down in a way that would sink a canal-boat, let alone a light canoe.

Several places during the day we saw the river out of its banks, and at a few points reaching from bluff to bluff. Game trails on the side hills were occasionally seen, but none of them comparatively fresh, the high-water having probably something to do with the scarcity. This is the land of the moose and caribou (the woodland reindeer of some authorities) the latter being decidedly the greater trail-making animal of the two, the former more of a rambler through the river bottoms and low valleys.

A characteristic of the Yukon River banks, throughout their whole length, is the thick, tough carpeting of moss on their sides, except at the most open places. This is so tenacious and hard to tear that in many places, where the swift current has cut out the bank at ordinary heights, the moss is left hanging over like a huge blanket, and trailing in the water below. So thick, tough and tenacious is it that on the lower river I have found it hanging down over banks full twenty feet high, and I believe it was strong enough to hold its own weight for probably double that distance; but as this moss is always rankest in the timber, the trees, in falling, as the bank is cut, tear off all the "moss blanket" below it, and these trees are not over twenty feet apart on an average. These pendant moss blankets were just commencing to show on the Newberry in a very few places that no doubt would have been greater but for the ravages of the high, swift water on the banks.

Floating very near one of the high sand bluffs, so as to be able to touch it with an oar, we could see that it owed its solidity to its frozen condition, the water having saturated it and the intense cold of an almost polar winter converting it into an Arctic sandstone, so to speak.

Just before camping that evening we passed several beautifully variegated bluffs on the east bank, probably old red sandstone predominating. From here we could also see distant blue mountains, probably twenty to twenty-five miles away, and stretching off in that continuous hazy aspect that generally indicates a distant stream, and, if so in this case, we surmised it to be the Yukon.

That evening one of the party asserted that he had seen a lynx near camp, or some large cat-like animal, that was hastily decamping. The lynx is the only one of the cat family in these northern woods, although I have heard a prospector, or miner, assert that there was a current rumor, generally believed, that a panther had been seen not far from the Forty Mile Creek Mines, near the Yukon River; but this is so very far north of the northern-most point they have ever been seen before that it is doubtful unless further corroborated.

The night of June 23rd–24th is chronicled in my diary as giving us the fewest mosquitoes we had yet had on the trip; and right here I might add that I found these torments in far less numbers than in my trip through the Yukon valley in 1883, when they passed a limit beyond the reach of righteous language. As Russell had passed two previous seasons in this Yukon River valley, and spoke of them as about equal to the present year, I was forced to believe that 1883 must have been a prize year for the pests. I feel sure I do not exaggerate when I state that I found a half-dozen in 1883 to every one in 1891.

The miners have the usual grotesque frontier exaggeration to express their ideas of this numerous insect. One declared that he found them so thick between the Yukon and Hotalingua rivers that he could make several saber cuts through the dense swarm with his bowie-knife, and holding it aloft, see the blood trickle from the blade, while a tobacco-chewing companion asserted that at one locality on the Yukon he would have to thrust his Winchester rifle-barrel into the mass and withdraw it quickly to get a place to spit. On the lower river Esquimau dogs have been killed by them, and it is asserted by some that bears, in crossing marshy districts swarming with them, have been known to stop and attempt to fight, rearing on their haunches bear-like, instead of retreating, and, after hours of combat, have their eyes so swollen with repeated attacks that they were closed, and often times wandered off, mired in the marsh, and starved to death. I have no bear-robes with me to verify the story, but I have heard it reiterated with such vehemence at different points in the valley, that it is not very hard to leave a few grains of belief in the mind of one who has fought his way through this country.

Again, the forenoon of the 24th, we heard a lynx scream in the woods of a heavily timbered valley leading off to the west, and showing that these animals were fairly numerous in these parts.

Early in the afternoon of that day, as we were slowly drifting down near the western bank, the boats close together, a moose was seen on the opposite bank browsing on the low, trailing, deciduous brush, and apparently unaware of our presence. The Winchester 40–82 rifle was in Russell's boat, the *Frederic Bonner*, and to him fell the risk of a shot at the animal from the turning boat, a good, honest one hundred and fifty to two hundred yards away. The movements in the boats attracted the attention of his mooseship, just as a low-branching cottonwood tree was brought between us by the boat's drifting and the slow walking of the animal itself. This prevented a fair shot at the very moment when the best shot at large game is nearly always presented, or just when an animal first notices his enemy and remains rigidly fixed and statuesque for a few seconds, seemingly uncertain what to do. The moose now stepped partially out from cover to satisfy its curiosity as to our identity, and the next second the rifle cracked, the animal plunging forward as if struck; but after several stumblings, due either to the effect of the shot or the tangled driftwood through which it plunged, it wheeled, and before another effective shot could be made, disappeared in the woods. From our point of view, we could see that the moose was on an island, and hoping that it had been hit hard enough to make it reluctant to take to the water beyond the island, we got out our oars in a jiffy and rowed across as only excited Nimrods can when the prize is a big one, losing less in downward drift than the width of the river, despite the swift current. It fell to Russell to finish his work, but after a long absence, with no shot heard on our own side, we felt sure he had failed, when we saw him returning in the distance. Upon rejoining us, he said he had seen the moose again at a great distance, running, but instead of one, there was a herd of three, and that none of them appeared to be injured in the least, all disappearing in a twinkle just as he fairly set eyes upon them. But the most surprising part of all to everybody was the fact that there was a dead two-year-old moose bull, just recently killed, about one hundred yards up the river from where the boats had stopped and Russell had first landed. It had been found by Russell, who, after seeing the others safely out of sight, had returned by way of the cottonwood tree to determine if he could find any traces to verify that he had wounded the animal shot at, and was utterly dumbfounded and correspondingly elated to find that he had a whole moose carcass to prove it—the animal having been instantly killed by a shot through the heart, the very part at which he had aimed. There seemed to be only one way to explain it. When the first shot was fired at the partially hidden moose it was killed instantly, but just as it dropped, a near comrade, that had not been seen at all, rushed into sight, and by dancing deliriously through the driftwood had led us to believe it was wounded; but after wheeling and joining the two others in the brush, had escaped as narrated. Russell's rifle had only stopped the moose, but his curiosity had secured us the meat, which otherwise would have been passed by and we would have chronicled this hunting incident as a failure. I once had a fairly similar incident happen while hunting elk on the South Loup, in central Nebraska, long before that part of the State was settled, about 1872 or '73, as near as I can now recall. I was hunting on horseback and with a small detachment of my cavalry company. In the chase of a small herd of elks I had become detached from my comrades, and was rather leisurely returning to camp, when I came unexpectedly upon another

Mark C. Russell with the moose he shot for camp meat, on the Teslin River, about one mile above its confluence with Yukon River, June 24, 1891. (U.S. Geological Survey photograph by C. W. Hayes) [Hayes #329]

but much larger herd of these fine game animals, and before they had seen me. My horse was a fine sorrel that would pick up the fleetest elk in a half-mile to a mile run on any reasonable ground; for it must be borne in mind that a broken country is in favor of these animals, and a level one against them as compared to a good running horse with staying qualities. As usual, I maneuvered around through valleys, down gullies and other low places to get as near as possible, which finally brought me to a steep, precipitous bank, and from its top I thought I would be able to begin the chase and close to my prey. My horse plunged up the steep bank, and I think I was as much surprised as the elk when I found myself directly in their midst—so very close, in fact, that they were breaking in a rapid run toward the river before I could get my Winchester from its boot on the pommel of the saddle, leaving behind them a cloud of dust that was the only indication of their course. My horse was too used to this work to be completely astonished out of his wits; so, before I had recovered sufficiently to give him a dig with the spurs, he had buried himself in the dust, and by the approaching thundering of the hoofs I knew we were gaining, when my steed stopped so suddenly that I thought he had collided with another herd coming the other way. I was more ambitious then than now, and not so easily deterred when I started on an undertaking, so I kept right on, going over the horse's head, and landed in front, my Winchester being the only part of the expedition that kept up with me. When I got the sand partially out of my eyes, I saw why my horse had changed his mind. In front of me was a precipitous bank seventy-five to one hundred feet high at an angle of sixty to seventy degrees to the horizontal, and down which the elk were rolling into the river in their bewildered efforts to escape. I pumped the fill of the Winchester magazine into the moving mass below, but was not particularly disappointed when the dust cleared away and I saw nothing, for it was about up to my expectations. After waiting here for a while, and noticing that there was no crepitus in my cervical vertebrae, I was about to resume my homeward journey, when a sergeant, attracted by my firing, came up. He was a fine hunter and had an eye like a hawk, while mine, in the then present state, would have to be compared to those of a sand-piper, with the sand predominating. He said he was glad that I had killed the elk, for he believed the others had secured none, and the fresh meat was out in camp. He also remarked that it was a good shot, for those from a height at moving objects are generally very uncertain. I refrained from surprise at his remarks, descended the steep bank with him, helped him to butcher a fine two-year-old doe, and, with a ham apiece on our horses, started for camp as unconcerned as if I had known it all the time. Her dull-brown sides, in the brown autumn brush, might have been a good test for Brazilian pebbles, but not for Nebraska sand.

This moose was the first large game Russell had ever killed, although he was a good shot and living for quite a while on the frontier, prospecting and mining. We took both hams and the choice bits, not forgetting the greatest delicacy of this animal to the Russians, who owned Alaska before us—the nose—and sailed away again, regretting that we were compelled to leave so much fine meat behind us.

In fifteen minutes afterward we floated out of the Newberry into the Yukon. The doctor's dead reckoning made the former 124 miles long from Ah'k-Klain, its valley picturesque and impressive throughout, while, in a more practical view, it was easily navigable for its whole length.

EIGHTH LETTER

The confluence of the Yukon and Newberry rivers is marked by a number of islands; and where about equally sized streams thus join, it generally leaves the impression that the one descended by the traveller is always the larger of the two. When I floated down the Yukon in 1883 the Newberry seemed the inferior, and now the reverse appeared more decidedly evident. Of course, the high stage of water in the latter and the low water in the former would emphasize this, for the relative stages were plainly seen in the muddy waters of one and the clear stream in the other—a relation which, if anything, was reversed in 1883, or, at least, normal. The Canadian survey instituted measurements later showing the Newberry to be the wider, but flowing less water at that date—a relation which was clearly reversed as to amount when we floated out of it. I think it quite likely that if a series of observations be maintained here throughout an average year, the Yukon would be found to be the lesser stream, but I would not be greatly surprised at the reverse. The great practical point is that the former, coupled with Ah'k-Klain, could bring continuous river navigation nearer the Pacific coast than could the latter, which is impeded by an impassable cascade and rapid not far above the confluence of the two.

A little over an hour and a half's drifting on the Yukon and a faint ascending column of smoke that arose beyond a low wooded point of land developed into a miners' camp, alongside of which we pitched our own, although it was yet early in the evening. There were two miners, "Cal" Swindler and George A. Harvey by name, who had taken up a placer bar that here projected into the river, but at that time so submerged that it could not be worked; so all were idle. They also had

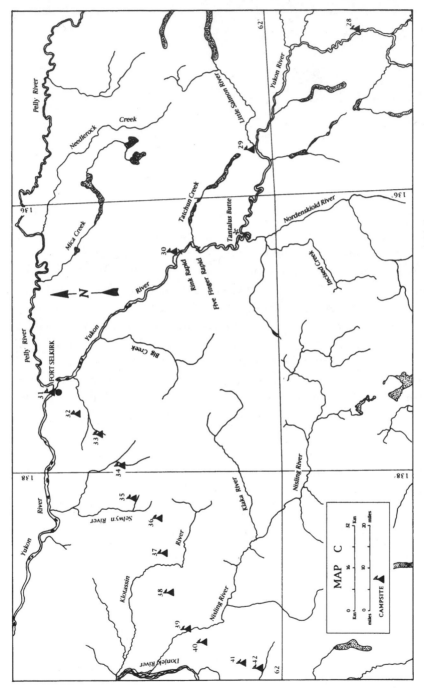

Map C: Route of the New York Ledger Expedition of 1891 from Camp 28 (June 24) to Camp 42 (July 20). Topography based Canadian Geological Survey Map Sheets: Lake Laberge (105E, Glen Lyon (105L), Aishihik (115H), Carmacks (115I), Kluane Lake (115G & 115F), Snag (115J & 115K)

a "Stick" Indian with them who was to go to work when the stage of water would allow. They expected a favorable fall in a week to ten days, but we afterward learned that it was much longer before they got to work, but were fortunate in making a good season's "run," after all. I will not tire my readers with a description of the placer miner's rocker and cradle system of working out an auriferous river bar, for that has been dwelt upon by scores of writers in the early mining histories of all the Western States and Territories. Swindler was a veteran miner in these regions, but it was Harvey's first visit to Alaskan fields; for though, strictly speaking, all of the placer mines of the Upper Yukon Valley are in the British Northwest Territory, yet the title "Alaskan" clings to them tenaciously, since the preponderance of prospectors are from there. Swindler had had an eventful prospecting tour the year before, extending across the whole face of Alaska, and debouching by the Kouskoquim River just in time to reach Juneau and "outfit" for a summer's work on this part of the river, an old "stamping ground" of his. His narrative was naturally interesting, although I recall but few of the salient points, the most conspicuous of which, in a sensational way, being the fact (or theory) that the notorious Chicago murderer, Tascott, was thought to be living among a band of savages on the Kouskoquim River, not far above the old Russian redoubt, Kolmakoff. The identification certainly seemed to be reasonably accurate, but too long to give in detail here without bringing this under "The Detective Series." It may be remembered by those of my readers having a surfeit of these sensational Tascott "finds," that the murderer was said to be traced to the Turtle Mountains of North Dakota and then lost in Canada, whose detectives finally traced him in a tourist jaunt toward Alaska, where he dropped out of sight.

We tendered our new friends some moose meat, but they only accepted a piece of the liver, they having a fresh supply, which they were just on the point of offering us. Swindler said moose were abundant in these parts, and that his Indian companion could guarantee him a carcass on a day's notice.

There are annually from one hundred to two hundred placer miners working on the bars of the Yukon River and its tributaries. From one-half to two-thirds of these winter near their workings, while the others depart each fall for the coast-towns and cities, returning the next spring. The migratory miners fit out in these coast places each spring,

Two miners, Swindler and Harvey, with their dog, Boodle, and their boat the Creeping Jesus, at their placer bar on the Yukon River, about twelve miles below Teslin River, June 25, 1891. Note the blurred image of their Native helper by the boat's bow. (U.S. Geological Survey photograph by C. W. Hayes) [Hayes #334]

generally in parties of two, or, at most, three, hire Indians to carry their effects over the coast range of mountains, and building a whip-sawed boat of a peculiar pattern, they descend to some bar they have agreed upon to work or prospect for another. The season over, they ascend the river by poling and tracking the boat along shore, abandon it at the foot of the mountains, and with but two or three days' short and correspondingly light supplies, they make forced marches over the range, arriving on the coast with a single blanket and their buckskin purses more or less full of gold-dust, according to their luck and work. Many, again, go "outside," as they call it, every second or third winter, and but few stay in permanently.

We sat around the camp-fire, "spinning yarns" until quite late, by comparison with our usual time of retirement, or as soon as our work was completed. Our lunch, afterward, was on moose's nose spitted on a stick and toasted over the coals of a camp-fire. It is of the consistency of gristle, with a nut-like flavor, not bad to the average palate, I should imagine. The Russians in Alaska were passionately fond of it, and all their old Indian trading stations, it is said by more than one writer, had a supply of these on hand for winter consumption, if there were any moose in the vicinity; and vicinity there meant anywhere within a radius of a thousand miles, if there was river navigation.

And this brings it vividly to my mind that I have eaten everything from beaver's tail to moose's nose, from warmed-over walrus hide to a ragout of rattlesnake; and that evening, for the hundredth time at least, was asked if I had ever encountered any gastronomic impossibility? I have.

A Minneconjou Sioux friend of mine once invited me to a dog-feast at the old Spotted Tail Agency, the fall of 1877. Now I can eat dog, if necessary, three times a day, but Touch-the-Clouds had sacrificed an old mesozoic mongrel, boiled it for a day or two in a weak molasses water and then sprinkled it with sugar to serve as a dessert. I ate all the Agency rations served and stuck flat-footed on the last.

We got away late that morning of the 25th, and in less than an hour's drifting, passed a deserted miner's cabin on the east bank. Here a couple of them had wintered a few seasons before. Another hour and we passed two bars pointed out by Russell, one Densmore's, on the west, and Hughes's on the other side, named, as expected, after their original discoverers and workers, and both now deserted.

Along the banks, at frequent intervals, were side rapids quite heavy enough to make us wish to avoid them. As they did not exist in 1883, when I drifted down this same part of the river, it shows how variable many characteristics of a stream can be, and how contradictory it may look to persons visiting at different times. Much of the conflict of description arises from this simple fact. A large number of these rapids on the Yukon and Newberry were of sufficient importance to be noted by a careful topographer, while another traveller, at a lower stage, would have failed to see them. In 1883 I passed two sets of conspicuously named rapids on the lower Yukon River, neither of which was visible to the naked eye at the time of my visit.

Ten o'clock saw us drifting past the Upper Cassiar Bar, as Russell informed me.[37] Here, in 1886, six men, in eighteen days, took out $800 apiece in gold-dust. As this is a fair average for a whole summer's work for two men apiece, on one of these bars, it can be seen by even the novice that the Upper Cassiar was very rich, but, despite which, it was deserted at the time of our visit. It is but twenty minutes' drift to Lower Cassiar Bar, and here we found one man in camp—Monahan—waiting for the water to go down. Fifteen minutes more, and we came to another temporarily deserted, its occupants, Dawson and Cummins, being at another bar a few hundred yards below, where Jones and Harrington were encamped, all, of course, waiting for lower water. As Russell and Cummins had formerly been partners in prospecting enterprises, an hour was spent here, when we got away, and drifted by the mouth of the big D'Abbadie River a few minutes later.[38] I named this river, in 1883, after the great French explorer, the local Indian name, translated, being the Salmon River, while the miners call it the Big Salmon River. It was not until we passed it in 1883 that any gold was found on the banks, although no great effort was made to discover it; so I announced the mouth of the D'Abbadie as the beginning of that precious metal on the Yukon, but the Newberry would have been much nearer the truth. The D'Abbadie, while noticeably smaller than the Yukon at their confluence, is, I think, the equal, if not superior, of the Newberry or Yukon at their junction.

It took less than an hour's drifting to bring us to another, and last, camp of miners, there being three of them—Brinton, King and Tremblé—on this bar, it, too, being out of sight under the water. Here we lunched, discussed the omnipresent and now tedious question of high water, and got underway.

That afternoon the moose's nose came to an end, as we ate it for a late lunch. Like all cartilaginous, nutty meats, it is, I think, much better cold than warm; but I agree with one miner who said: "I don't want too much tree-topper snout in mine."

---

37. Upper Cassiar Bar was one of several gold-bearing bars found in the summer of 1884 and named by Thomas Boswell, Howard Franklin and Michael Hess, prospectors who had previously mined in the Cassiar country of northern British Columbia. The bars produced the first considerable amount of gold mined on the Yukon (Lewes) River (Coutts 1980, p. 50).

38. The D'Abbadie River (335 river miles above Dawson City), named by Schwatka in 1883, is shown on recent maps as the Big Salmon River (Coutts 1980, pp. 23, 71).

Just before going into camp we passed a "Stick" grave, the totem-pole alongside of which was about as good a representation of an ordinary barber's-pole as one would expect such barbarians to make who had never seen one. It surmounted the crest of a high bank on the east side in a most conspicuous way, and made us all feel of our rough faces and wish that we could go and get shaved.

We camped near the mouth of a river that I named the Daly in 1883, and which the Indians call the "No Salmon," in contradistinction to the Salmon (D'Abbadie), which, by comparison, has a greater abundance of these fine fish, the miners calling it the Little Salmon.[39] This was one of the most beautiful camps we made on the river, the moss on the ground being soft, yet not wet, while the timber was more open and penetrable than usual, many birds singing in the woods, and a few pretty squirrels chattering and running around through the brush. From a high hill near by, a great many lakes could be seen inland, the incoming river here being evidently a mere chain of them a little farther inland.

A very heavy dew fell the night of the 25th–26th, the first one we had noticed, and almost equal to a light rain.

That curious insect, the "devil's darning needle," I noticed was very numerous that forenoon. Some of my readers may know that some scientists have advocated the breeding of this creature, if it can be done, as a mosquito-destroyer. This would be a good country for the test. I was much inclined to believe that the high water had much to do with the failure of the mosquito supply—much more than the "darning-needles," at least—and that as soon as it fell and stagnant pools were left standing everywhere, the mosquito census would soon show its prevailing percentage. These pools are much more liable to be left by the receding waters in this country than those of lower latitudes, since in the latter they may drain directly through the soil, which is generally pervious, while in the interior of the Northwest there is a perpetual impervious stratum of frozen soil a few feet below the surface that forces the water to drain laterally into streams or remain as pools and puddles. This undrained condition of the sub-Arctic countries is, after all, the main cause of the dense swarms of these pests that populate it.

39. Schwatka named the river (200 river miles above Dawson City) to honor Charles P. Daly, President of the American Geographical Society. The miner's name, Little Salmon River, remains in use (Coutts 1980, p. 161).

That very forenoon we had ample proof of this. Drifting by a high bank that had recently been cut out by the high, swollen river, we could see where the frozen sub-soil began, and along the sharp line of demarkation the water was draining out in volumes, at one point spouting almost clear of the frozen bank.

The Eagle's Nest, a prominent broken butte on the river, which I named in 1883 from the local Indian appellation, was passed early.[40] The eagle's nest which gives it its name is still there, perched on a beetling pinnacle that lies deep in a great fissure of the cliff.

The Yukon is very winding at this part of its course, no less than seven reaches of the river pointing toward a single prominent hill, Tantalus Butte, before it breaks away from it.[41]

So far, all of the large tributaries of the main stream had come from the east, but about noon that day we passed the mouth of the Nordenskjöld River flowing in from the west, a stream I named in 1883 after the celebrated Swedish Arctic explorer.[42] Near here beautiful rosebush patches were seen on the sloping banks, while in one or two places on the level bottom-land wild hay could be seen growing profusely that looked much like timothy, and that was nearly as high as one's waist.

About the middle of the afternoon I became somewhat anxious regarding the Rink Rapids,[43] that I knew to be ahead a short distance, for the high stage of water might make it somewhat dangerous to shoot them in light canvas boats. As a usual thing, rapids that flow *over* obstructions are improved in navigability by high water, while those that flow *through* them are only rendered worse. The Rink Rapids are of the latter class, a number of huge rock pinnacles projecting from the very bed of the river, and looking for all the world like the piers of a great

---

40. The prominent landmark is now known as Eagle's Nest Bluff. The worst riverboat disaster to happen on the Yukon occurred here on September 25, 1906, when the *Columbian* was destroyed by an explosion, killing six men (Coutts 1980, p. 88).

41. The name Tantalus Butte for the lone hill five miles north of Carmacks, was given by Schwatka in 1883 and remains in use today (Coutts 1980, p. 260).

42. The name Nordenskjöld River (258 river miles above Dawson City) is still in use. (Coutts 1980, p. 197). The present village of Carmacks and the Klondike Highway bridge over the Yukon River are located one mile above its mouth.

43. Schwatka's name, "Rink Rapid" (242 river miles above Dawson City) was never in common usage, and its earlier and more popular name Five Finger Rapid is now used. In later years a fixed cable was installed to allow river steamers to be winched upstream through the rapids. (Coutts 1980. pp. 98, 223).

stone bridge whose spans had tumbled in. I think there are five of these at least. Some miners speak of it as "The Five Fingers," while the Indian name is unpronounceable. About three o'clock we entered very rough water, extending from bank to bank in a contracted part of the river, and from the commotion I thought we were close to the rapids, although the map showed them some distance farther on. It took about an hour's swift drifting to reach the Rink Rapids.

When I shot these on a raft in 1883, at the request of my Indian guide I beached our craft and made an inspection beforehand, choosing the extreme right-hand channel, the usual one taken by the Indians and miners. The excitement then was just pleasantly exhilarating. An inspection of the rapids by Russell and the doctor gave no reason for changing the usual channel, or method, and at 4:20 we went through, Russell's boat slightly leading.

The right channel is, after all, quite a narrow chute, and the greatest danger is in not keeping well to the center, to avoid collision with the rough, jagged rocks of the sides. At the entrance the first wave is about four feet high, and they increase in size to about the third or fourth, some two feet higher, with well-defined "combings" that are fully equal to the heaviest surf one usually sees on the Atlantic shore, and whose hissing roar makes it hard to hear anything else in the narrow cañon. The importance of a little headway for steerage power in such places cannot be over-estimated, and the doctor and I dug in our paddles to secure it, the consequence being that we collided with the other boat in a sharp rap, that might have sent it to the bottom had it been an ordinary wooden affair. This was just as Russell's boat mounted the wave ahead, and the next second he disappeared out of sight in the trough beyond. As we rushed into the white-cap aloft, we could see that our prow, projecting beyond the comb, was directly over his boat, and it seemed certain that it would come down on top of it as he rose and we sank. A couple of quick back-strokes with our paddles, and no further collision occurred. Russell, entering alone, had to rely upon his oars, which are poor enough in heavy, rolling seas, and came near being thrown sidewise on a breaker's crest and swamped; an accident which I think not unlikely the collision helped to avert.

In any dangerous chute, a boat of our size should have at least two persons aboard, and paddles relied upon wholly. When we shot these, in 1883, the waves were not quite so high, and the waters seemed more broken up and frothy, the whole length of the short cañon being a mass of milk-like foam, while at this stage of water the huge waves, or, better speaking, great, white-capped breakers, were well defined, reaching from rock to rock, and did not break into a boiling mass until the cañon was cleared and there was a chance to spread. One of the singular sensations of shooting an exciting rapid is the apparent slowness with which one moves while directly in the worst and what one would infer to be the swiftest parts, from the very name, "rapids." It always seemed to me that the boat was very deliberate, and rather liked dallying in the dangerous waves; and then, again, it may be true that the foaming, dancing, roaring water is not going as fast as the senses credit. I know that is true of similar people.

Little less than an hour after we came to a rapid that was something of a surprise to me.[44] I had seen nothing of it on my former visit, and though marked on maps made after, I felt a little skeptical as to its amounting to much, to say the least, until I came in front of it that afternoon, when my skepticism received a very rude rebuff, for these rapids were as formidable looking, at first sight, as those we had just shot, and they seemed to reach from bank to bank. Fortunately this was only apparent, for an inspection revealed that by hugging the east bank we would escape the worst of it, and this was accordingly done.

The roar of these rapids could be heard for half a mile away, and in the most violent parts sent sprays of water aloft like spouting geysers. The waves we had to ride were quite high, but long, and not crested with curling white-caps; so, outside of possible sea-sickness, if they continued too long, there was no great danger, and we really enjoyed them, by comparison with the much more dangerous breakers in the Rink Rapids. This experience proved to us conclusively that these folding canvas boats were superior to wooden ones, leaving out the consideration of portability, where it is evident, and which the novice would assume to be the only point wherein they were superior. Outside of poling up-stream along the shallow shore, I know of no place where I would not prefer the King folding boat to any wooden one of equal capacity. No canvas boat is stiff and rigid enough to stand the warping strain of two men poling it fore and aft.

---

44. This rapid (230 river miles above Dawson City) has been known officially as Rink Rapid since 1887 when George M. Dawson transferred Schwatka's name. Rocks that formed the main obstacle to steamboat passage were blasted out in 1902–1903 (Coutts 1980, p. 223).

Singularly enough, Russell did not know that the collision took place while in the Rink Rapids, although it was a very severe one, as he was so busily engaged in righting his swinging boat to escape swamping. It was more like the collision of two foot-balls than two cannon-balls, as with wooden boats, the shocks being almost wanting.

A most furious wind had made the steering most aggravating, and, of course, at the worst places. It had amounted to a gale just before shooting the Rink, and at one time we discussed the advisability of postponing the run until it should die down. Just after, we were treated to the incongruous combination of a prolonged heavy rain-storm with the sun shining the whole time, and half of that square in our faces.

The white volcanic ash stripe had followed us down the river, and just below the lower rapids we saw a bank of it in a deep curve of the bank that must have been five or six feet thick, the deepest deposit we saw.

We camped early, satisfied with the variety of exciting adventures, if the distance had not been very great.

As I stretched out beneath a clump of poplars,[45] I mused on the extensive habitat of this tree on the North American continent—the American cottonwood, as it is usually called. I have traced it from Panama through Mexico and our own land to where the biting boreal blasts of the Arctic Sea render all tree-life impossible, and it holds its own with every species that flourishes in this wide extent.

Where the *alamo* or the "cottonwood" will not take root and grow, there the reindeer-moss and the scurvy-grass of the Arctic had better beware, and the cactus, mesquite and sage-brush of lower latitudes had better look out. It is as omnipresent as the poor, and as versatile as a Yankee inventor, and ought to be the national flower if American characteristics were the test.

The next morning (June 27th) was so cold and cloudy that we got under way early for exercise. The cutting capacity of the swift Yukon River was well seen in a way that forenoon when we passed the old Indian village of Kit'l-ah-gon, or rather its site. Here a large, fairly well constructed building stood in 1883, with a large number of brush-houses around it, all on a level but low bank that reached far back to

---

45. This interior species is Balsam poplar (*Populus balsamifera* L.). (Viereck & Little 1986, pp. 72–73.).

the hills, and that was freely covered with the everlasting poplars. Now the river had swept away every vestige of the native town and cut deep into the cottonwood grove, leaving only a frowsy fringe of it hanging over the bank. In this part of the river were many Indian graves, and their abundance may account for the want of a native village any longer in these parts. Some of these I recognized as old acquaintances of 1883, but most of them were new and more recent arrivals—or rather departures.

Shortly after noon we sighted the Pelly Cliffs, a prominent line or bank of basalt which showed where the Pelly River came in. This latter stream is an old, old highway of the Hudson Bay Company and used by them considerably when they occupied this country many years ago. It is now abandoned by them.

A half-hour later we saw the site of old Fort Selkirk, our objective point for one part of the expedition, and in a short while were alongside it.

NINTH LETTER

F ort Selkirk[46] is an old abandoned station of the Hudson Bay
Company—one that had been deserted for nearly a half-century,
so long ago was this ancient company of Prince Rupert's found
feeling its way far out into the frontier wherever furs of any kind could
be obtained. I think it was along about the early 'fifties that they established this place, so far away from their base of supplies in Hudson's
Bay, that it took them two years to reach it then. They found themselves poaching on the peltry preserves of the Chilkat T'linkit traders of
the coast, who, then, outfitted by the Russians, were doing a good
business. The Indians know but one method of competition in business. They went into no intricate inventory for the purpose of reducing
stock, nor did they place more flaming advertisements before their
customers. They simply organized a war party, rapidly descended the
river, briefly burned the buildings and appropriated the goods. At the
time of my first visit, in 1883, there were still partially standing the
three stone chimneys of the building, but in 1891 these were reduced
to three stone heaps that looked more like the ruins of a prehistoric
race than anything civilized.

But while this old landmark of civilization had almost completely
disappeared, another had sprung up directly alongside of it in a fine
large log-house, built by Mr. Harper, an American trader. When I first

---

46. Fort Selkirk, located 178 river miles above Dawson City, was founded by
Robert Campbell in 1848 and abandoned four years later after it was burned by a
party of Chilkats. The log buildings described by Schwatka were built by Arthur Harper
in 1889 when he opened a trading post on the site (Coutts, 1980, pp. 102–103).
Whatever remains of these and the more recent structures dating from steamboat
days are being refurbished by Parks Canada. A useful bibliography is available
(Dobrowolsky, 1988).

Log buildings and fenced garden, Fort Selkirk, Yukon, July 8, 1891. (U.S. Geological Survey photograph by C. W. Hayes) [Hayes #308]

floated down the river, the farthest inland station on the great stream was at Nuklakayet, about six hundred miles from the mouth, and Mr. Harper was in charge of it. Now Selkirk was on "the ragged edge of the frontier," and Mr. Harper had, like a true frontiersman, moved along with it. I had hoped to find him here, as his influence and help would have materially assisted us in the further prosecution of our journey. Our original plan had contemplated, in fact, that our real explorations would begin at this point, or possibly from some point on the White River, and bear from this as a base toward the southwest. A straight line from here to connect with previous explorations on the Copper or Atna River, or any of its eastern tributaries, would probably give the longest line that could be laid down on the American continent wholly through unexplored country. How we came to put in a preliminary exploration from the mouth of the Takou River, near Juneau, to that of the Newberry, appears in the first article.

Mr. Harper had left some two or three weeks before, taking his furs down the river to sell them at St. Michael's, near its mouth, and procure other trading material. The station was in charge of an Indian from farther down the river, who introduced himself as "Sam." He said he belonged to the Takudh band, generally spoken of, I believe, as the Tadoosh. He was a good-natured fellow, and understood a little English, but his interpretation was occasionally of rather an aggravating order. He always assumed to understand anything that was said to him; and as oftentimes he did not, the result was not satisfactory. An inquirer having confidence in his interpreter is easily misled by the latter's "yes" and "no," which are interjected as assurances of understanding; but Sam's were so often at variance with the questions asked that it did not take long to place this fault—really one of excessive good nature—at its true value. A long-sustained mental effort with Indians is more tiring and irritating than any physical exertion, and for this reason an Indian interpreter is not so good as a white man, if both are equally conversant with the languages used.

There is another quite interesting difficulty in interpretation with savage languages that I think is more the fault of the English tongue than due to the savage's construction, and one which will often put the novice completely off his conversational base if he is not on the watch for it. I refer to the opposite popular understanding of a negative question. To illustrate. You ask both a white man and an Indian the question: "Are you not going to cross the river?" and both equally answer: "Yes." A little while afterward you find the white man on the other side of the stream and the Indian still remaining on this, the "yes" in the two cases having directly opposite meanings. In the above simple case it is easy to see wherein the two interpretations have differed, but if one will reflect that there are often long-winded, intricate inquiries, wherein the meaningless negative element is lost sight of by the English inquirer before he is half-way through it, while the savage gives it full credit and replies accordingly, it is not difficult to understand why gross errors can arise when the answer thereby means directly the opposite of that intended to be conveyed. A spade is a spade, and a negative is a negative with Indians, and there is no circumlocution of linguistics that will twist their language into any other construction. While two negatives do not always make an affirmative, still one of them always passes at par value at their palavers.

There were also two local Indian boys with Sam, who were engaged in whip-sawing rough lumber from a log, at the time of our arrival. These were all the Indians in sight, so the chances of getting about a dozen to act as porters and packers for us to the Copper River did not seem very flattering, coupled with Mr. Harper's absence and the lack of a good interpreter. Where others could be found was not so apparent either. Sam and the two boys thought that a raft-load or two might be coming down the Yukon or Pelly within two or three weeks, and seemed to act as if we would be overjoyed at this information. After a council of war, that would have been the fortune of a "dialect" author, I induced the two boys to go to drum up recruits for my expedition, or rather to induce all able-bodied Indians that could be found anywhere to visit Fort Selkirk to confer with me and have a feast. I did not have any faith in the conference as an inducement, but I knew they would come in from Hudson's Bay on crutches for the other. I simply mentioned the conference, so that my proposition would not startle them into a choking fit at the feast. The various articles of trading material I offered had less effect in starting them than the promise of a dozen "hard-tack," or pilot-bread crackers, which were given as rations for the trip. They got away the middle of the afternoon of June 27th, so after a little work pitching camp, there was nothing to do but await developments.

I tried to take a bath in the river; but, though its waters were high, its temperature was not, so my revelings were short. Cleanliness may be next to godliness in the temperate zone, but the remarks usually made in securing it in the sub-Arctic regions are not of that character.

The middle of the afternoon next day, Sam informed me that a raft could be seen floating out of the Pelly River, and that he believed it held some Indians that probably had been sent in by our messengers. They turned out to be our two couriers themselves. Their report was not any too flattering. They said they had found a camp on the Pelly, but it contained only squaws, and that they did not care to go packing; and for which I did not blame them, as I had not asked them. Some other Indians had been seen, and they had reported they would not miss the feast. One fact looked ominous. They had killed a moose and said they did not want to pack. I did not exactly see any reasonable relevancy between the two, but they evidently did.

About an hour or two after this report came in, another raft was seen floating out. It had on board some six or seven Indians who with a vast number of dogs were soon alongside. It was about the most forlorn and decrepit crowd of savages I ever saw—if one can conscientiously apply the word "savage" to such a meek mass of humanity. There was not a man among them fit for packing, and they looked more like an ambulance load destined for a hospital than a party from which any work whatever could be expected. And, sure enough, their greatest

A group of interior Natives from the Pelly River in front of a log building at Fort Selkirk, July 8, 1891. (U.S. Geological Survey photograph by C. W. Hayes) [Hayes #326]

demand was for medicine, both for external and internal application, and it never ceased while we were with them. The most convenient ingredient we had to spare was carbolated vaseline, and we doled it out in homeopathic quantities for everything, from corns to consumption, and from baldness to bunions—all of which it cured. I never saw a better example of faith-cure in all my vast backwoods medical practice. If I could impress one-tenth the confidence in it at home, this equivalent of Colonel Seller's famous eye-water scheme would become a grand and rich realization.

In the way of advice, as to the routes for reaching the far-off Copper River, and maps that explained them, they were as prolific as possible; and after accepting the first installment, there were avalanches of advice, it rained routes, and maps—almost by millions—fell like flakes in a snow storm. Fifty men could not have kept track of all, but one sleepy boy could have attended to all the applications for work. The advice was mixed as well as massive; for as soon as one borrowed a pencil and paper and began on a map, all the others got around and began a Babel of boisterous explanations and corrections, that ended in more of a lung-power test than anything more rational. Their lungs and larynxes were the only perfectly healthy parts of their anatomy.

One curious piece of information arose partially as a result of imperfect interpretation and partly from the direction we desired to go—that is to the Copper River; it was that these Indians knew more or less directly of a copper deposit, or mine of this metal, somewhere on or near this river. Every time we spoke of Copper River, they diverged to the copper deposit, as if the former was a sore subject to them, for some reason, or that the latter was one on which they liked to dwell. They seemed so persistent, and as I finally believed that the two places were so near each other I gave up any other idea, except that of the latter, reasoning that if it could be attained the other would be close at hand, as a result of so doing. There was one hardened old sinner among them, who had been given the American cognomen of "Jackson," and who felt so proud of it that he would not give his Indian name. He professed to know more than the remainder of the natives about the copper, and intimated that he knew the Atna chief, Nicolai, or Sko-ti, as he called him. He said that it was eighteen sleeps to the copper deposit, and but one from there to Nicolai's house; but it might be well to state here, that he varied as to the number of sleeps with each map he drew, and he was topographically very energetic. The variations as to the copper might possibly have been reconciled by assuming an equal number of scattered deposits, but it was very hard to believe that Nicolai was so ubiquitous and miscellaneous, or had such a large number of houses as were needed to fulfill the other discrepancies in distances.

TENTH LETTER

Among the men of the party I noticed that most of them had their ears and noses pierced as receptacles for the usual savage ornaments, but their utterly absurd and senseless ear-piercings were generally about a half-inch to an inch above the usual place in the lobule affected with so much grace and appropriateness by civilized beings. In fact, the piercing is almost directly behind the aural orifice, and produces a most singular effect when conspicuous earrings are worn. Most of them were without them, however, and the majority of the boys were without piercings, from which I inferred the practice is dying out.

There was no doubt about there being a copper deposit somewhere near by, which was visited occasionally by these people, for we found among them copper bullets, copper arrow-heads, a long copper spatula for digging the marrow out of caribou and moose bones, solid chunks of copper not yet beaten into any shape, and a small buckskin bag about half full of finely powdered azurite, or the blue carbonate of copper. The latter was used as a pigment by the ladies for ornamenting their cheeks and chins by the tattooing process—another fashion which, I believe, is slowly dying out, very slowly, but still dying out a little faster than the tribe itself.

The change from the Copper River to the copper deposit did not render everything clear sailing, however. I soon ran into a superstition that was used by them with irritating effect occasionally. They allege that if one of them strikes a bowlder of copper with an ax—or any other instrument, for that matter—he will die soon after. In vain I told them that I did not want them to strike the copper. In some mysterious way, that good interpretation might have cleared, they seemed to be impressed with the idea that there was great bodily danger risked at the copper mine. I knew there was great financial danger in fooling around any sort of a mine, but I had never heard this specific charge before.

Early the next forenoon, Jackson informed me he would readily drum up any Indians in the vicinity if I would "outfit" him with a good supply of white man's food for the journey, which I did. In the evening he returned with the cheering information that he had been following a fresh moose trail all day, and had forgotten all about the Indians. I suppose he wanted some meat to go with his bread.

I got the two Indian boys to start again that evening to look up more (?) packers, and again turned in to wait for something to turn up.

The dismal prospect for packers had prompted some to discuss the feasibility of getting some squaws, should they be necessary to fill out a quota. The standing of women in the different savage tribes of my acquaintance varies through quite wide limits. Among the Eskimo they are treated about equivalent to the women of work-people with us— that is, they do the household work and attend to the children mainly, and the men do the rest; while among most Indians the women are practically slaves, doing all the work, except that attendant on the chase and war. The Pacific-coast Indians of Alaska, however, are a conspicuous exception, the women ruling the property, the lines of regal descent, and in many other ways have rights that are unknown even in our own civilization. A man with more than one wife is often seen, but women with more than one husband are not unknown, as an offset.

These Indians among whom we found ourselves leaned more toward the true American Indian than the Alaskan coast tribes just spoken of. A good incident that occurred illustrates it better than any general description. Two miners of the Yukon valley had married squaws, one (we will call him Smith) having secured his bride from the coast, the other's (Jones's) sweetheart having been an inland maiden. Their summer's work over, Smith and Jones started out together in two boats up the river to reach the coast. My informant passed them at a point on the stream where there was good "tracking," and in the first boat was Mrs. Smith, steering the craft, while Mr. Smith had the tracking-rope over his shoulder, tugging along the bank; while just back of him was Jones stretched out, half asleep, steering his boat, and Mrs. Jones playing the part of the tow-path mule.

Thunder-showers are almost unknown on the lower Yukon River, but I have already spoken of their comparative frequency on the upper

A group of interior Natives in camp near Fort Selkirk, July 8, 1891. (U.S. Geological Survey photograph by C. W. Hayes) [Hayes #328]

tributaries. There was hardly a day during our two weeks at Selkirk that a thunder-shower did not pass over or near us. The heaviest of these seem to pass down the White River valley to our westward.

Early in the afternoon of next day, a large raft was seen floating out of the Pelly River, soon afterward followed by two others not as large. Later, we could see that the forward raft was made up of two smaller ones lashed together. They were covered with Indians and dogs, but did not seem disposed to stop, as none could be seen working at the oars, and they were drifting down quietly on the other side of the wide river. They were evidently surprised at seeing so many Indians on the shore near Mr. Harper's house, and the usual animated conversation sprang up across the waters, and led to their rowing the rafts ashore on our side, landing nearly a mile below us. There were from fifteen to twenty Indians in the whole party, and our prospects looked more hospitable-like and less hospital-like than they had. Late in the evening I unfolded the tender subject of packing over the trail, though I got no enthusiastic volunteers, but a great deal of information about the trail. I determined now to wait a day or two and let them study it over. An idea that has to pierce an Indian's cranium not only has to get through a thick skull and tough hide, but also a dense thicket of hair; so it must be given time, unless shot in with a rifle.

Native camp near Fort Selkirk, Yukon, July 8, 1891. (U.S. Geological Survey photograph by C. W. Hayes) [Hayes #331]

Among the thunder-showers that appeared on July 1st was one to the south, with a black, funnel-shaped cloud, not unlike those delineated so often in illustrations of cyclones, although I greatly doubted if it indicated any such energetic meterological phenomenon.

The 2nd, I again tried to do something definite toward getting packers, but had little success. They spoke of a few more expected in, and then said they would decide. In a sort of indefinite way, I ascertained that I might get four packers, possibly five, out of those already camped near us. It was late in the day when we heard a shot in the mouth of the Pelly, and to which our Indians replied with compound interest, and soon after a small raft came in sight, bearing a man and two nearly grown boys. The former disdained to even talk about packing, and we afterward learned he was a sort of a sub-chief in a small village far down the river. The boys were not so unwilling, but they were equally far from willing.

Two more families got in the 3rd, but matters stood about as usual. Five seemed to be the limit I could depend upon. This number was not flattering, as a test of their carrying capacity also revealed, as we had expected, that they could not compare with the T'linkit packers in any way. While the T'linkits could easily average one hundred pounds over the severest mountain trails, these fellows would do well to carry sixty apiece. Some of the party suggested packing the dogs. We had seen a number of these used for that purpose in moving their scattered camps

Three Native boys, Iz-yum, Mug-ich-luk , ("Harper"), and Peter, at
Fort Selkirk, June 28, 1891. Hayes noted, "Peter...has no Indian
name as the missionary caught him very young and baptized him."
(U.S. Geological Survey photograph by C. W. Hayes) [Hayes #333]

together, and here they staggered along under thirty-five to forty pounds,
as near as we could estimate; but what they could do on the trail was
not so clear. It was mournfully evident that the white men would have
to do a fair share of the work if we ever expected to make a start. The
aversion to this was not based on laziness at all; but a map-maker,
photographer, investigator of any of the natural sciences or chronicler
is more than handicapped, who has to carry his share of effects and
attend to these duties while keeping along with a lot of racing savages,
who have no more idea of the importance of these than they have of
cosmology.

Independence Day came. I was greatly surprised to find the savages
had long anticipated it, and were longing to celebrate it. I might have
expected it in Sam, who came from the Alaskan side, lower on the
river; but where the others caught it, the good Lord only knows. Here,
on the Queen's sacred soil, the very savages were turning themselves
inside out celebrating another country's separation from their sover-
eign, while inquiries showed that her majesty's birthday was unknown
to them. I can recall but one case that is in any way similar to it, and

here the cause was radically opposite. I was in San Francisco one summer, when a bluff British tar brought in a steamer-load of Chinese. He was fined outrageously for carrying passengers in excess of lawful limits, imprisoned for a contempt in not reporting it, and drawn and quartered for some other technical delinquency, despite which he made a display of bunting on the Fourth that outvied anything American in sight. His answer to an inquiry was: "I was celebrating the day that England got rid of this blarsted country!"

I think the Indians' desire to celebrate was based on an omnipresent eagerness to burn gunpowder and make noise, and that the patriotism in it, like the result of the explosions, was all smoke. Guns and pistols were banging away at frequent intervals, and in the evening several tough stumps and logs were blown up in lieu of a national salute. I noticed one instance of the economy of physical exertion that rather impressed me. The problem was to bore a hole in a two-hundred-pound stump with a two-pound auger. Two Indians carried the stump about a hundred yards to the auger, bored the hole, then carried the stump back to where they found it, and blew it up in honor of the American eagle. If the American eagle could not manage any better, he would need the motto that now reposes between his uplifted wings.

The Indians of the Yukon River love to waste powder on every occasion that will give them the least pretext. Every one who comes in sight up or down the river in any sort of a craft is generally thus saluted. In 1883, when my party was drifting past the village of Noo-klak-o, the nearest trading station upon which they could depend for powder being over eight hundred miles away, they stood upon the bank and ground out a revolving reception of about one hundred and fifty shots, until I could hardly tell whether I was drifting by a Chinese New Year's or a Gattling-gun proving-ground. I had a box of one thousand rounds of ammunition with me, but I kept the temperature of my gun-barrels very much lower than my excitable entertainers, and which they seemed to resent to an extent that a liberal gift of trading material could hardly appease.

The Indians offered us many things for sale, and, while most of them were of a trifling character, still there were some I would like to have had for various reasons; but I knew well that if their longings for money or material were satisfied with tradings, that the chances of getting them as laborers were correspondingly decreased.

We were surfeited with information about Münchausen mines of every character, and I suppose it was clearly beyond the comprehension of Indian intellect for any one to come into such a country without some such object as prospecting or mining, exploration being away beyond their mental grasp and understanding. It is a fair basis on which to meet the savage, in that he will comprehend you more readily, and, last but not least, he might lead you to something of value in that line, although the instances on record are far from numerous. All this, of course, if the two paths followed are practically the same, and the main object is secured in apparently following another, which, in itself, would be a beneficial discovery if realized. The most profuse individual in these matters was Jackson, and he talked about mines of fabulous richness, until I thought I could hear Ananias roll over in his grave and groan. There was nothing small about his mineral estimates. His pieces of precious metals were not as big as one's head or fists, the usual extreme limit of frontier exaggeration, but they were as large as log-houses and moose-skin tents, and I believe the eternal hills themselves would have been drawn on for comparisons, but even Jackson knew a hill had to cover a deposit to make a mine out of it and the mathematical axiom that a part cannot be greater than the whole. He placed one deposit of copper only a day's journey north of us carrying packs, but was so earnest in his desires to be paid a dollar or two for the information before starting, instead of twenty on the ground after verification as promised, that no one was deluded into further consideration of it, beyond talking about it to pass away the time.

It now became irritatingly lonesome with so little to do, and no definite time ahead when it might end. There were a couple of miners' boats here that had been traded by the owners for supplies, so it was said, they expecting to winter or leave the country by way of the mouth of the river, as was occasionally done. Sam claimed one of them as his property, so I made conditional arrangements for its purchase, and we started to repair it. My intention was that, in case sufficient packers could not be procured here, to make the trip from Selkirk to the nearest Copper River tributaries, to take a couple of Indians in this boat and with the party descend the Yukon to the White River, then pole and track up it to some one of the Indian villages of which we had heard and there procure packers for the remainder of the trip. It is always a safe basis to assume that Indians not in close contact with white men are easier to procure for small exploring parties and more faithful in their work than those who are, although this statement does not imply, as one would infer at first glance, that the white men are at least indirectly, if not directly, responsible for it. In nearly every case they are not, but all the reasons are too long and uninteresting to dwell upon here. It is this fact that has made clear the explanation of the phenomenal success of the earliest explorers in wild countries where later investigators met such unwelcome receptions, resulting, oftentimes, in utter lack of success.

These boats of the Yukon River miners are patterned somewhat after the *bateaux* of the Canadian *voyageurs* of the British Northwest Territory, and yet there are some distinct differences. The *bateau* is made for larger parties and is consequently larger, two persons usually forming a mining or prospecting party with the former people, while the *bateau* is only limited in its size by the weight that can practically be transported across portages and around other obstacles. For equal weights the *bateau* will carry the greater loads, while for equal size the miner's boat is much stronger, stiffer and more adapted for poling up stream against the rapid currents of this country. To compare the American miner and Canadian *voyageur* as boatmen is a harder task. Each one is undoubtedly superior to the other in his respective vocation; but for general, "all around" boating, such as is needed in exploration, I believe I would rather trust the miner (be he from Canada or the

United States) as against the *voyageur*. In this I mean the miners of this region, who have to do a great deal of boating in their vocations, and achieve a wonderful degree of skill.

Even in preparing for this boat trip it was hoped that it would not have to be undertaken; as, after all, it simply postponed the outfitting for the main trip until the villages were reached, and imposed a preliminary undertaking as irksome and nearly as formidable as the main one.

About noon, July 5th, a miner's boat was reported coming up the Yukon, and shortly afterward it was alongside, the occupants being a white man and two Indians from the placer mining region on "Forty Mile" Creek, a western tributary of the Yukon, coming in something over two hundred miles below Selkirk. Mr. Frank G. H. Bowker, a young American miner of British birth, was in charge, and had two fine-looking Indians with him, that made the Selkirk natives look more contemptible than ever by comparison. They were from the same country as Sam, while they rejoiced in the full-length titles of David and Peter, and resented the curtailment to vulgar "Dave" and "Pete." Both spoke better English than Sam and negotiations thereafter with the others were more easily carried on. This party was going westward, prospecting; and as Bowker had full authority from Mr. Harper as to Indians, supplies and other necessaries, it became easy to unite the parties, to the advantage of both. The moral effect on the local Indians was quite apparent. I got six of them as packers at once, with the promise of a half-dozen dogs, and the assurance that the latter could carry thirty-five to forty pounds apiece, especially if the load were of that character that diminished as the time advanced.

I very much preferred this starting from Selkirk for the reasons given, and the further fact that it was over a trail I had heard of in 1883, and one of the most important ones of this northwest country. It was made, so the Indians then told me, when Selkirk was first established, and was used by the distant natives on the head of the Tanana River, as well as all lying between. It was an extremely interesting country, so I had been informed; and as nearly all explorations in the far northwest had so far been along the principal water-courses, I knew that one "across lots" ought to give considerable novel information.

During the leisure time that fell on our hands awaiting the action of the natives, we often talked with them about outside matters as well as we could, through Sam's interpretation. The latter had some books, and tried hard to learn English, having been taught to read by missionaries lower on the river, when a boy. He showed me the printed version of the old familiar hymn, "Happy Land," in the Takudh tongue, and as it may prove interesting to some of my readers, I transcribe it:

I.
Tseyiu nunh kug kooli
Nizhit, nizhit,
Zut rsyotitinyoonut
Negichilzi
Thlih ha zyunchigahudit
Ako dhundei ungitli,
Kreist vih ekwahudit tsut,
Shegungitli.

II.
Ei ssokooli nunh kug,
Jesus kwitchin
Zut Kreist ttsun nilinut
Kookwutechya
Thlih ha sso kwitelya
Vittekwichanchyo chinttsi
Kwikeitrutunahtyah
Sheg, sheg kenjit.

The Indians around camp were constantly singing, or, more correctly speaking, humming some tune, and most of these could be traced directly to the common favorites of miners and traders, such as: "The Arkansas Traveller," "Rory O'Moore," "The Girl I Left Behind Me," and similar airs, that are usually the first frontier attempts to replace the savage tom-tom and wild war-whoop by a very slight improvement. Occasionally they sang in concert some air that had evidently been taught to one of them, years before, by some missionary, and the effect was really pleasing. They are instinctively musical, but their voices need cultivation.

Mr. Bowker gave us much interesting information about the mining region around "Forty-Mile" Creek, probably the most isolated mining camp in the world. Here news is received but once a year, and then of a meager character. Bowker said that, when poling up the Yukon, the year before, he inquired from a newcomer who was floating down if the United States had yet elected the President. The reply was that a "new feller named Harrington, or something like that," had been selected.

They had quite a late season that year in getting to work, owing to a large "tidal" wave sweeping down the river and carrying all improvements before it. It had been estimated by some to be thirty feet in height, and was probably caused by the sudden bursting of an ice gorge on the upper stream or one of its large tributaries.

There were two tame moose in the mining camp, that had been domesticated by the miners, who, as a class, are usually quite fond of wild-animal pets. These deer carry bells on their necks, and it is well they do, for they often absent themselves for long distances from the camp, and would have been killed long before but for this precaution. One of their favorite jaunts was to swim the broad Yukon itself, for which they seemed to care no more than for the smallest stream. Whenever they walked down the single path in the mining-camp that did duty for a main street, everything else politely gave them the right of way. A small bull-terrier was the only opposition to this emolument they ever had, and that was short. The terrier had a playful way, that greatly attached him to the miners, of running out from his master's cabin at any passers-by and inserting his teeth into their calves or other distal extremities. He tried it on the bull moose-calf, one afternoon, while spectators were near. The moose was grazing slowly down the path; near the cabin the dog rushed out and started for his hind legs, when he got more of them than he really intended, the bull-calf letting both feet fly as if the terrier had touched the trigger. The moose actually did not stop his grazing until he heard the dog drop a little later, when he looked around, stared at him for a while, then yawned and began grazing again. The dog, stupefied, returned the stare without a whimper, then slunk away, and was never known to bite a man or a moose after that. The moose was not quite full-grown, but he was as much larger than the dog than the United States is larger than Chili.

TWELFTH LETTER

The country is one that is unusually well stocked with large game of the sub-Arctic *cervidae* family, and it is rare that moose or caribou meat cannot be had in the winter's mining-camp for from ten to fifteen cents a pound, and occasionally as low as five cents. When "hard-tack" crackers are five cents apiece, this makes the figures on game ridiculously low. Bowker, like most young Englishmen on the frontier, was fond of field sports, and did considerable hunting. On the "bald divide," or high tracts barren of timber, lying between what the miners call "Forty Mile" and "Seventy Mile" rivers, he says he has seen herds of migrating caribou, probably two thousand and three thousand in number, while on every side were these animals making the snow fairly gray. This was in the late fall when the cold snaps of approaching winter were determining the migrations of these vast swarms. At this time a single Indian, with a Winchester magazine rifle, killed one hundred and five of these animals at a single run, often killing two at a shot with this hard hitter, and a number to which he limited himself only by the fortunate fact that he ran out of ammunition. This meat was mostly sold to the miners at five cents a pound, they buying large quantities, as it could then be kept through the winter.

Bowker's best story was, fortunately, about fish. These were caught on "Forty Mile" Creek, where they must be superlatively abundant. A miner there had to live on them until he was sick of fish. It was nothing to catch them, and he simply walked down to the perpetually set net and took out what he wanted, liberating the others. Killing some game one day, he gladly pulled the net out of the stream up to the bank to keep from catching the fish, but that night the river rose, covered his net, and when he visited it next morning it was as full of fish as ever.

One thing is sure, I believe, and that is the Yukon River, with its tributaries, is the greatest fish stream, for its size, in the world; and it is the seventeenth on our globe, the seventh of the Western Hemisphere, the fourth of North America, and the third of the United States. If it has a rival, it its either the Mackenzie or Back's Great Fish River of British America, but neither of these is in any way so accessible, and Back's is likely to remain inaccessible.

One of the comical incidents of hunting had happened the winter before. There were five actors in the scene—two hunters, a moose cow and her calf and a big timber-wolf. The plot shows that the wolf wanted the calf, to which the mother moose objected, and that the hunters wanted everything in sight—wolf, calf and cow. The hunters, while separated, were yet close enough to talk to each other without alarming the animals, so deeply interested were the latter in each other, but, singularly enough, so disposed was the timber that one hunter saw only the wolf while the other saw only the cow and calf, and neither knew of the other's information, but supposed they were stalking common game. After considerable sportsman's strategy, they were ready to fire, and the moose-hunter whispered to the wolf-hunter to take the "little one" as his shot; the latter, thinking, of course, he saw only one of several wolves, as they nearly always run in packs. At the word the guns went off, both bringing down their prey; but the moose-hunter seeing the calf run away, the resulting conversation was about as follows—at least the non-sectarian part of it:

"You missed the calf, the best meat of the two."

"That's the first time I knew a wolf ever had a calf, or that wolf-meat was worth eating; but—" here the moose-calf came circling back to its dead mother—"what in the name of holy hairpins is that?"

The answer to which was a shot from his companion's Winchester that mortally wounded the animal, that went stumbling backward, pursued by the moose-hunter, who nearly stepped on the dead wolf before he saw it, and who, thinking the wolf was about to spring, went up into the air with a spring that would have done credit to an Apache. It took considerable time to straighten out the plot, even to the actors, or at least those that were left of them.

Nearly all of the Indians near these mines work on the bars in the summer, and in the winter make a good income by hunting for the whites who remain there. Many of these middle Yukon River Indians will not eat any breakfast before going on an important hunt after moose or caribou, or one requiring long-continued physical exertion, claiming that it is injurious for them to do so and jeopardizes the undertaking. This is somewhat in contrast with the T'linkit Indians, who insist on three full meals a day, if procurable, whether hunting, packing or doing other hard work. In general, savages fill to repletion when they have it, but can reduce rations to low ebb, or do entirely without them, with less grumbling or other apparent effect than whites; but there are exceptions and great variations to all these statements, as seen above. Even among different classes of frontiersmen this variation is noticeably apparent, and some occupations will be actually "stampeded" at the prospect of food running short before another class will begin taking it into consideration. In the same white classes the variations of nationality are also often apparent.

I have spoken of one petty sub-chief from lower on the river, who so disdainfully refused to pack for the party. Bowker's Indians knew him well, and a good story was told on him. He was often independent to the point of insolence in his dealings with the miners, and this class of people are not given to taking this when strong enough not to endanger the camp by resenting it; so when Lo, the poor Indian, one day, intimated that a miner had fractured the truth, the knight of the pick and pan prospected his face for a fissure vein, and ended by tossing him into the river. The average American miner is one of the biggest-hearted men living, and, when he has actually got a foe beaten or cornered, one of the most forgiving; so Lo was almost instantly pulled out, and his point conceded about the misunderstanding as to the moose-meat bargain, the miner not being angered at the Indian's demand, but at the insolent way it was put in questioning his veracity. Next day, however, Lo put in an appearance at the miner's cabin with a fine, large moose-ham; but the white man could not understand what for; so, sending for an Indian interpreter, he got the following explanation: "He say he want to give you this moose for saving he life yesterday when you pull him out of river." It need never be said that the Indian is wholly without gratitude.

The 7th of July was one wholly of business, and the next day was appointed for the start, reserving the right and expecting, however, to postpone another day if the start could not be made early.

We procured nine dogs for the trip, and set some of the squaws to work fixing their pack-saddles so as to render them waterproof. These are simply two enormous pouches, one on each side, holding about

A Native cache including sleds and snow shoes in a spruce
tree along the trail. (U.S. Geological Survey photograph by
C. W. Hayes) [Hayes #330]

three gallons each, attached to lashings, neck and waist-bands that are needed to fasten them to the somewhat mobile skin of the dog's back. Outside of this packing, these curs are eminently worthless in the summer time except as scavengers for the refuse [of] decaying salmon; but in the winter season they are used to draw the native sledges or toboggans and to assist in trailing moose and caribou. They are half starved in the summer and the other half is what doctors call "low diet." Everything eatable must be kept away from them; so, near each camp or village, one will see the lower branches of a tree utilized as *caches* for provisions, skin-clothing and even toboggans and snow-shoes, if they have their raw-hide lashings on. I do not think, however, they are as bad as Esquimau dogs. I have known the latter to eat or tear to pieces sole-leather, pistol-holsters (kindly leaving the pistol), canvas gun-covers, cloth saturated with grease, tarred rope, and sheathing of skin-canoes and, *en route* to Back's River from Hudson's Bay in 1879, had them tear to pieces a pair of India-rubber overshoes that I was depending upon for summer wear on King William's Land.

The Esquimau method, in use on the lower Yukon, of harnessing dogs to their canoes or boats like canal horses and towing them along the banks, I did not see in operation during my stay among these people at either visit, although they possessed all the requisites for such an easy and convenient method of navigation.

Our first solicitude as to our new pack animals was to feed them, but any attempts to fill them up was like the proverbial pouring of water down a rat-hole. In the struggle for food, we were greatly annoyed by the raids of the other dogs to secure their share, these Indians having none of the ingenious devices of the Esquimau for this purpose, besides having a much greater unwillingness to do anything in the way of work they could avoid.

Every matter progressed favorably on the 8th, and it looked as if we would get an early start next day, but those who have seen much field work know that first-day starts are seldom early.

In wandering around near the camp of some Indians I visited to look about packing affairs, I noticed a small patch of wild strawberries that gave me a few ripe specimens, but hardly up to the standard of the wild ones we see at home. The Pacific slope of Alaska is far more prolific in this fruit, and in many places it is much more palatable than were these representatives. Singularly enough, the finest-flavored and most abundant strawberries were found in the ice-clad and glacial regions, a sort of Alpine variety, no doubt. In my Alaskan expedition of 1886, these were found quite abundant in the Mount St. Elias region, especially around Yakutat and icy bays.

One of the small birds I noticed near this strawberry patch was the omnipresent "whisky jack," the most ubiquitous winged thing in the northwest. Wherever man, savage or civilized, is found in this region, there the "camp-robber," as the miner calls him, will be found. The "Hudson Bay bummer" is another unsavory title given by the frontier people he meets in his own country. It is almost impossible to expose anything around camp of an eatable nature if one of these birds is near. It is a trifle smaller than a jay, which it closely resembles, except that its color is lighter. I have known it to light on persons in camp if they were asleep or otherwise quite still, so great is its familiarity. I never heard so noiseless a bird, as it flits through the brush making no more noise than a butterfly, until a sudden flutter is made so near the head that one is half frightened out of one's wits at the intrusion. The most comical case I ever knew was where a hunter had killed a grouse and put it in the forks of a tree just above his head, as he was asleep on his bed. A "Hudson Bay bummer" came flitting along and started to feast on the dead grouse, when the man awoke, and, seizing a convenient club, hurled it upward at the energetic bird. He had forgotten about the grouse, so when it came tumbling down in his face and the "camp-robber" sailed away unseen, he thought it was an attack by this audacious bird, and seized it excitedly and violently beat the ground with it before he discovered his mistake.

THIRTEENTH LETTER

The last article left us talking about the camp birds, or "whisky jacks," of the Northwest Territory. It has been said by some scientists that there is considerable mystery about their nesting habits, and oölogical specimens from their species command high prices; in fact, one institution had offered a goodly sum for a perfect nest of this bird, with its eggs intact. In connection with this, I heard from a miner in this region that they nest unusually early, building under the snow alongside of bogs and brush, as he had seen the young being fed by the mothers before the snow had hardly started off the ground, or about April.

It was high noon on July 9th before we got away, it having taken from very early in the morning until that time to bring the various loads of the men and dogs into a state of equilibrium, and the comparative mental satisfaction of the bearers.

The first hour's walk was over a level tract of hard ground, the bottom land of the wide valley, that was bordering on the luxurious for travel, by comparison with the Takou trail. In that hour I believe we made the equivalent of some of our severest days' labors on the latter. The white men were now carrying from thirty to forty pounds on their backs, and I must say that it appeared a mere bagatelle to me by comparison with half that amount before, so different was the walking. Whenever we came to groves of spruce and other conifers, they were open and park-like in their distribution, with but little underbrush instead of a dense mass of darkened timber interlaced with a terrible tangle of brush and all of it accompanied with moss, or a morass, into which one sank at every step, or scrambling over boulders[47] and fallen

---

47. Previously spelled "bowlders."

logs, forcing one into devious paths to avoid them that almost double the length of the trail. There were many pretty flowers in bloom on all sides, and in every way the country seemed more like the woods and prairies at home than any part of the far Northwest that I had ever seen. After this hour's walk on the level flat, we came to the low rolling foot-hills where the trail was quite as good as on the level. The only trouble was the slight one of water being very scarce, and the day being quite warm. The Indians found a miserable little mudhole off the trail; and, although the water was cool, it was so strongly impregnated with the taste of leaves, wood and other forest *débris* that our lunch, taken at this point, was not a very palatable one, despite the conversion of the black water into blacker tea.

In the valleys of the rolling hills, great groves of quaking aspens[48] were found, and here the trail was unusually good, the ground being generally carpeted with a soft, firm turf from which sprang a luxuriant growth of grass. I know nothing of the severity of the winters in this region, but unless they are too cold for stock to live here during that season, there is certainly good reason why they should do fairly well. Of course, stock-raising on a large enough scale as the sole occupation, would be out of the question for want of a market; but a person resid-ing permanently at any point could have a few cows that would make living much more enjoyable than without them. Game, however, is too plentiful to make meat production any great consideration.

Among the groves of poplars and aspens one could often see the "girdlings" of the trees by the hares gnawing the bark for food, and sometimes to an extent that killed the tree. The various ages of the different gnawings—for I think they are made only when snow is on the ground, and these little animals are thereby cut off from grasses and herbs—were apparent, and I suppose indicate the depth of the snow during different winter seasons. If this is so, this depth is not very great at any time—certainly not enough to interfere much with stock-raising, while the woods would give ample protection in the coldest winds. Numbers of these hares or rabbits are secured by the Indians; and some Indian families, especially those deprived of the support of good moose and caribou hunters, often depend upon these smaller animals when fish are not to be obtained from the Yukon and its tributaries. We

---

48. Quaking aspen or trembling aspen (*Populus tremuloides* Michx.). (Viereck & Little 1986, p. 76–77).

had such a family with us at Selkirk, the father being dead, and the widow and two small boys of ten and twelve years being indefatigable and very successful in the chase of these northern hares. These hares are much better eating, I believe, than any of their species of lower latitudes; but the Indians say that at certain seasons they are not so good, owing to parasitic worms, at a time, usually, when fish, and especially salmon, are abundant. Even at these times they kill and eat them if they are on the march away from the fish-bearing streams.

There were many burnt timber-tracts passed by us that day, but, as in the case of all open timber interspersed with many little prairies, these tracts were small. The maximum injury by fire to an unsettled or thinly settled region is reached in either a densely timbered country or one perfectly open, as our great Western plains. In a happy mean between the two, the trees are not close enough to assist the conflagration without the help of a strong wind, while the grass is not high, dense or dry enough to aid it in that way. Aquilegias, sometimes but erroneously called wild honeysuckles, were not infrequent along the trail, while some very pretty orchid-like flowers were often seen.[49]

I have spoken frequently of the "trail" as if this was the usually well-defined Indian path leading from one region to another; but in this case there were some limitations that might not be uninteresting to describe. As usual in open countries where walking is equally good over wide areas an Indian trail often scatters to an extent that no defined path can be seen at all; and the natives simply follow a general direction, hunting, berry-picking or what not, on the way. Where a constricted part of a valley or a pass through the mountains is met, there the scattered parts may unite so as to be more noticeable. Through a densely wooded or rough mountainous country, full of impediments to the pedestrian, an Indian trail is always much more plainly marked for the same amount of travel; yet it is one of the most singular facts in the world, to me at least, that not one tribe in a dozen will ever attempt to improve such a trail. A dozen to a score of Indian packers, with great loads on their backs, will stoop to their very knees under a vine maple or leaning pine branch, week in and week out, going and coming, when a single blow of a small ax would sever it, and a vigorous thrust throw it aside. I can partially understand the selfishness of an Indian going over a new trail, as with our T'linkit packers from Juneau; but where a

---

49. Aquilegia, or Columbine (*Aquilegia* L.) (Hultén 1968, p. 457).

trail is periodically passed over by a band, as with the interior Takous on the same path, it is perfectly incomprehensible. In fact there is no essential difference between an Indian trail and a game trail of migratory animals, and I have seen well versed frontiersmen study as to which kind a certain path could be. Twelve times out of a dozen, I believe, an opinion on either side would be correct, and the trail is used indifferently by the game or human animal, the only difference being the Indians have a sort of right of way in using it whenever they want, while game will not travel it while the scent of the former is yet clinging to it from a recent trip. This is especially true among the timid animals, as the woodland *cervidae*, while bears and the *carnivorae* are more aggressive.

Late in the afternoon we came to rather a singular natural phenomenon. After crossing a very warm, marshy stream, the waters seemingly too tepid to be accounted for by even the sultry weather, we came to a series of pocket-like wells on the mossy banks, where the water was as cold as ice. These wells, or rather long cylindrical holes, would often hold only one's arm, and it would have to be inserted to its full length to reach the water, and one was lucky in getting a cupful of clear fluid in the shallow bottom. After the first cupful it was generally impossible to obtain more.

A very few grouse were seen, Peter killing one. Hudson Bay "tea" was unusually prolific here, and we came to a profusion of light-colored moss that the natives and frontiersmen here call caribou moss, according to Bowker's testimony. A characteristic of this caribou moss is in its constant moistness, whatever may be the state of the weather; and to this is attributed most of its palatableness to the animal from which it derives its name.

Just before camping, we passed an Indian fox-trap made of poles. It was of the "dead-fall" variety, that killed the animal when sprung.

It was half-past eight before we went into camp; but as far as weariness was concerned, I felt as if I could have walked the rest of the night, had it not evidently been wiser to spend the time in sleep. In fact, that evening, while thinking over some matters, I found myself walking up and down the open space near where my bed had been made down, but this was a casual recreation that was not repeated for many days later. So far, the dogs had kept along nobly with their thirty to forty-pound loads; but in the late evening, we began to appreciate what a nuisance they were, after they were let loose and began rummaging around after something to eat, smelling around the head of the bed and occasionally, as a diversion, fighting over the foot of it.

The country was quite open, while directly ahead of us, bearing south-southwestward, as a sailor would say, was a conspicuous hill from two thousand to two thousand five hundred feet in height.

That night mosquitoes were much more numerous than they had been along the river, auguring ill for our future comfort on this part of the trip. Even worse than the mosquitoes, however, was the singing of one of my packers, who was a whole light opera in himself, his *reper-toire* beginning and ending in "The Celtic Laundress," and which he kept humming far into the night, but not far from my bed. The humming of the mosquitoes was much less irritating.

It was late when we got away next day, but the weather looked auspicious. Some delay was experienced in the bad interpretation of the Indians about some food supplies. Their demands were perfectly reasonable, but, as interpreted, it looked as if they wanted Alaska and a mortgage on Canada, and it required some diplomatic strategy to get matters straightened all around.

An hour or so before noon we came to a long, pretty slope, facing the south and trending off toward a stream, and its sides, in places, were fairly covered with strawberries. Here the whole party scattered to gather some of the luscious fruit as they passed along. At the same time a bunch of spruce grouse was flushed in the bottom near by, but they were so very wild and wary that none were secured. There is no doubt that the wildness of game is generally rendered greater by the constant hunting of them, but again there is often found a wariness about them that cannot be accounted for on any such basis, as in the present case. These Indians seldom hunt birds and we were the first white men ever over this trail, according to the native accounts. Even with some of the larger kinds of game I have seen this apparent anomaly hold good.

FOURTEENTH LETTER

R eaching the Strawberry Creek (it sounds, perhaps, comical for the far northwest under the very edge of the arctic circle), we saw a few grayling in it, and some few of them were caught by the white men with a colored fly, the Indians getting some in a less artistic manner with clubs and poles.

The warm afternoon gave us a very refreshing head breeze, the whole benefit of which we got in this open, rolling hill-land. It was the same old southern summer breeze.

There were a great many tolerably fresh moose signs seen that day, but we were never destined to come in contact with that noble game animal again on this trip. In the way of small game, Jackson, who had come along as guide, secured five "snow-shoe" rabbits, or northern hares, the former name being given by the whites of the country, on account of the broad expansion of their pedal extremities in the winter, when the hairy covering is at its fullest development and its impressions on the soft snow are enormous for the size of the animal.

Whenever we came to a deep creek, a bridge of logs would have to be built to enable the dogs to cross without wetting their loads. These deep creeks were not infrequent the first few days out from the Yukon River, but soon disappeared and gave way to wider, shallower crossings. Even here the dogs have to be watched, for if the day was warm and they were fatigued, they would lie down in the water to cool off; and as many were inclined to loiter, it can readily be seen what a nuisance they were to superintend. That day, July 10th, was particularly bad on the dogs, on account of the fallen timber that lay nearly everywhere across the trail. These were mostly small logs and did not interfere greatly with the men, but for that reason were greater obstacles to the canine packers than larger timber would have been.

The evening's camp was near a clear stream, and a fair mess of grayling were caught as a consequence. These fine game fish are in every clear stream on the head of the Yukon that has yet been investigated, and without any reference to the muddiness of the receiving stream through which they must have passed to reach the clear tributaries. I have seen grayling in streams so small that one could step over them from bank to bank, and so shallow that they must have been frozen solid every winter; yet these clear-water rills flowed directly into the muddiest glacial streams that one could imagine, and where one would hardly believe that any fish could exist, and that a game fish would especially abhor.

We now began encountering an obstacle to pedestrianism that is not uncommon all over interior Alaska and British Northwest Territory; I refer to a sort of coarse bunch grass that affects marshy places, which, when dry and black, shrivels up into clumps…with just enough room to put the foot between occasionally…. Of course there are many variations in these obstacles with the common characteristics of always appearing infinitely worse than any of the varieties met so far. Shortly after noon, next day, we reached the top of a high ridge, or "hog-back" of far western parlance, eighteen hundred feet above our morning's camp. It was dry and timberless, but water could be had by scrambling down to the little wooded pockets that put up the ravines. Some of the Indians took a short cut on thirst, by stripping the bark from birch-trees and appropriating the wet sap beneath. Far ahead, and apparently none above our own level, snow banks could be seen on hillsides and ridges. The afternoon's journey can be condensed into moose-brush and mosquitoes.

Late in the afternoon, the Indians pointed out the blue line of the White River.[50] By sunset we had seen so many very fresh caribou, moose and bear signs, that we felt confident and hopeful that some of those animals would soon be encountered, as the dogs greatly needed several good feedings. One of the unarmed Indian guides well ahead came back and begged a pistol from Bowker, as he explained this country was one famous for savage bears, and an encounter might be expected any time, while plunging through the high brush. We encountered no

---

50. This was the eastern branch of White River, now known as the Donjek River. The name Donjek was applied by Hayes, apparently from the native word, "Donyak" or "Donchek" (Coutts 1980, p. 86).

bears and lost no Indians. Soon after this, Jackson kept the game excitement up by reporting that he and his boy had shot a caribou, but that it had escaped. We certainly heard a shot fired, but Jackson was so reliably unreliable that there seemed to be some doubt about it.

By seven in the evening, we had made our way down from the ridge into the valley, and shortly after came to a small stream that was undoubtedly a headwater branch of the Selwyn, a stream I had named in 1883 while drifting by its mouth, and after the director of the Canadian Geological Survey.[51] Its waters were clear and full of grayling. An hour later, we reached the main Selwyn, ascended it a short distance and went into camp. Bowker prospected on a "rim-rock" ledge of the stream, but only got a light color or two for his pains, David fishing and catching a fine mess of grayling meanwhile. The Indians were quite confident in their predictions that a caribou would be killed on the morrow, near the white snowdrifts seen on the ridges ahead.

Leaving camp next morning, we ascended the Selwyn a short distance and then took up one of its westerly branches. In this latter, the grayling were so abundant that the Indians hastily extemporized spears of forked poles, and caught quite a number of them. Just before noon, we had come to a number of snow-drifts in the creek-bottom, and shortly afterward began climbing the ridge, on which we found them more abundant. We had got well over this ridge and were descending into the valley beyond, when a shot was heard far ahead, and, on coming up, found that David had killed a fine two-year-old caribou cow. Its butchering disclosed that these people, besides rejecting the parts we also reject, in addition avoid the liver and kidneys, except in cases of dearth, feeding all to the dogs; in fact, the heart is the only interior organ they use. All around us were other fresh signs, with a few indications of wild goats, also, while plover were numerous on the high, barren ridges. We got off the high ridge into the timber lower down, and went into camp, where wood and water were abundant. That evening, as we went into camp, we came on mosquitoes in the greatest abundance we had yet seen on the expedition or that we ever saw afterward. It was impossible to breathe without getting them in the mouth and nostrils, while a few holes in the top of my half-worn-out hat were such vulnerable places of assemblage for them, where they crawled in like bees at a hive entrance, that I wore out the crown completely in

51. Schwatka's name Selwyn is still in use (Coutts 1980, p. 238).

beating it to keep them away. That day, on the trail, a hornet or wasp stung me severely, and in a general sort of a way I wished I was back in the land of rattlesnakes, centipedes and tarantulas.

During that day and for several days afterward, the tops of many of the ridges were serrated with projecting pinnacles and pillars of rocks, that gave the rolling hills and otherwise almost pastoral scenery an extremely picturesque appearance. I had noticed this castellated appearance of the high rolling ridges on the Yukon hills near the mouth of the White River, in my former visit to Alaska in 1883, but little thought it was so very common farther up the valley of the latter stream as it now showed. I had thought the stream near which we camped that night was a branch of the White River, as it trended so to the northwest as far as the eye could reach; but my Indians assured me it was a tributary of the Selwyn, and, after flowing northwest, emptied into a northeastward-flowing branch of this stream.

Near us a lake had been dammed by an enormous ice-gorge, not half of which had yet melted, the stream now flowing underneath. This half-covered icy lake, with the snowdrifts here and there on the ridges, despite the timber on the lower levels, gave the country a more generally arctic appearance for a July aspect than I ever saw for the same month on King William's Land, in the arctic region, where Sir John Franklin's large party had so miserably perished with cold and hunger.

Next morning we awoke to the fact that the dogs had eaten up most of the caribou during the night, and, though it was not lost to us by this theft, yet, like half-starved brutes, they had over-gorged themselves until they were very loth to move, whining piteously when their masters beat them rather than move along.

From the morning's camp, the prospect ahead had shown high hills to cross in reaching the White River basin, but we were most agreeably surprised in finding the trail led through a very low, winding pass, that had been concealed by the overlapping hills. That night we camped on a tributary of the White, a basin we were destined to keep until we should debouch into the valley of the Copper.

The next day, the 14th, we made some thirteen miles, a little over the daily average we had been travelling so far.

The forenoon of the 15th gave us some pretty high climbing, and in a few hours we were twenty-two hundred feet, by our aneroids, above the morning's camp. From this ridge we could see the valley of the main White River, from which ascended a number of scattered smokes, some of which, our guides told us, located the camping-places of other Indians, while others, as we could see, were caused by spreading forest fires, probably started by accident. From here we could also see snow-clad mountains to the far southwest, our first glimpse of the distant St. Elias Alps or some of their interior-bearing spurs. Bearing south by west, twenty-five to thirty miles away, was an enormous lake,[52] at least twenty miles in length, but the air was so hazy with smoke that estimates alone were necessarily wild and uncertain.

Late in the afternoon, while the party was strolling along almost in a bunch, gabbling like geese, we popped over a ridge that disclosed a two-year-old caribou bull in full sight, about seventy-five to one hundred yards ahead, Bowker getting the first shot and securing it. It was in better condition than the first one, and that night we reveled in caribou steaks, liver and boiled tongue, the dogs, too, getting a fair share for their trouble in carrying most of it into camp.

---

52. Kluane Lake, the largest lake in Yukon Territory. The name was first recorded by Professor Aurel Krause of the Bremen Geographical Society in 1882 (Coutts,1980, pp. 149–150).

FIFTEENTH LETTER

T he next morning after killing the caribou was an extremely warm one, and we did not get started until late. That afternoon we passed through about a section of bottom-land covered with as fine a quality of wild hay as I have ever seen. It was waist-high in places, and would average two tons to the acre, I believe. Shortly after, we came to a series of small lakes, the water very shallow, warm, full of skippers and smelling badly. Digging through the reindeer-moss near their flat shores, ice could be found six to eight inches below the surface. We were now within a couple of miles of the smoke on the main tributary of the White and on the banks of a large branch of it. Two signal-shots fired by us were answered from the river beyond, and in a short while we were at our first Indian camp. This stream is fifty to sixty yards in width, the water about middle deep, fairly swift (four- to five-mile current) and fordable, with good rock-bottom, the Indians calling it "Ripple River."[53] There were native camps on both sides, the main one, of fifteen to twenty souls, being on the opposite bank; for which reason we camped on the nearer shore. Here the spruce-trees were the tallest, straightest and largest of any grove we saw on the trip. We were now just ninety miles from Selkirk, by the doctor's dead reckoning, and half-way to the Copper River, or copper deposit, by the Indians' estimates.

The communal fish-trap had not yet been built, as the salmon were just beginning to ascend; but in a day or two they had it done, and some fine fish were caught. The natives here say the Ripple River empties into the main White about ten to fifteen miles farther down, and that on the head of it, about a week's foot-journey (packing) from here

53. On his map of the Yukon District—Sheet 2, (Hayes 1892, Plate 20) Hayes recorded this as the Nisling River. The name remains on recent maps.

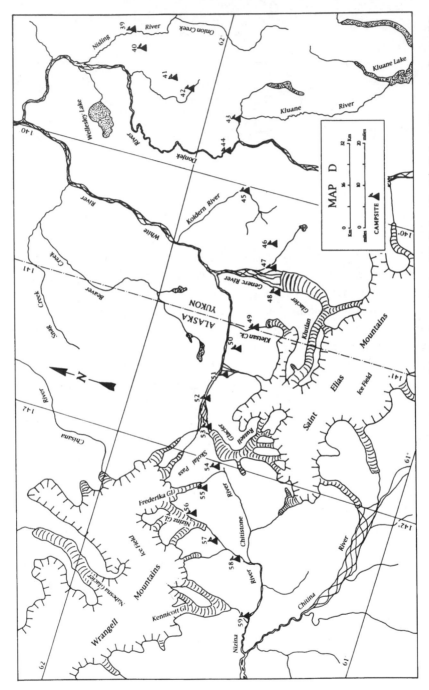

Map D: Route of the New York Ledger Expedition of 1891 from Camp 39 (July 16–17) Yukon River, to Camp 59 (August 10). Topography based on U.S Geological Survey Alaska Topographic Series: Bering Glacier, McCarthy, Nabesna, and Canadian Geological Survey Map Sheets: Kluane Lake (115G, & 115F), Snag (115J & 115K).

(sixty to ninety miles) there is a large lake that takes five days to go its circumference in a canoe, or one day across—about twenty miles, as near as I could ascertain, for a modest and probable under-estimate. A prospect of the riverbanks here showed only a few light colors.

Fortunately, our new friends were fairly supplied with dried moose and caribou meat, even if salmon was yet scarce, as we purposed making this a refitting point for provision and an exchange of packing-dogs, if possible, with probably a new Indian packer or two. There were many wild currants near this camp, but even when ripe, as they were, so sour they puckered up the trail if they grew near it.[54] So there were long intervals between pickings after the first mess; in fact, the second picking has not yet taken place.

These Indians had many specimens of copper, but, instead of having been made into utensils, were mostly in roughly hammered pieces or native masses of a few pounds each. I left Selkirk, as I have before explained, without any definite arrangements as to my Indians, packing to the mythical copper deposit or Copper River, hoping to get assistance from some of these White River villages clear through. I now learned from these natives that they considered the latter trip impossible in the summer-time, and almost so in the winter, even when the ice covered the numerous roaring torrents and snow filled in the crevasses of the many glaciers that had to be encountered in the rough mountain-pass. In case we could not break through the St. Elias Alps, there was still left us the way out to hug their eastern slopes to Chilkat, Alaska, and pass over much unexplored ground; but the original plan of reaching the Copper was far from being given up. None of the Indians here, however, would even think of discussing such a trip. We had now "shown our hands" so plainly to all natives, by these discussions, that we were determined to attempt to break through the, to them, impassable mountain barrier, that they raised all sorts of objections to prevent it; no doubt thinking so far that we would be compelled to return with them at any point they saw fit to turn back. They urged that they would be held responsible by other white men for our loss; and, while this idea seems trivial at first sight, it should be remembered that the savage mind does not discriminate so closely between the murder

---

54. Northern Black Currant (*Ribes hudsonianum* Richards.). (Viereck & Little 1986, pp. 152–153).

of a white man outright—for which the frontier sometimes holds them too rigidly responsible—and their loss by carelessness on their part, or even, as in this case, should we never be heard of again, by the responsibility of the whites themselves. Their customs as to corporeal responsibilities are often at such variance with our own, that their deductions seem as absurd to us as ours probably do to them. Then I think there was something in the less commendable reason of wanting to have the return trip in full wages for packing us back. They wanted to be paid every night after camping, if they went further, which was clearly impossible, and in many other ways too intricate and tedious to dwell upon; they tried to drive this idea out of our heads, until it seemed as if the very sword of Swear-ocles hung over the expedition.

On the 17th and 18th (July) we refitted, exchanging for some fresh dogs, and getting two additional Indian packers, and, the middle of the afternoon of the latter day, were glad beyond measure to get under way again and leave behind the all-day sulkings and machinations of an Indian town, sandwiched in between restless nights of medicine-men howling like panthers and thrumming of tom-toms, and the whole gamut of savage rites, superstitions, music and others "too numerous to mention." One of the new packers was a young man of rather stupid appearance, whose father was one of the grizzled veterans of the camp; and just as we started, the old man threw his arms over me, begging me to look after his boy, and rubbed his rough beard against my chin until I looked about twice as stupid as his son.

I noticed in this village that the auxiliary sign-language, or that which is profuse in gestures and facial expressions as accompanying explanations to words, was quite common; but whether any code of the pure or pantomimic sign-language, so common with many of our Indians on the plains, is ever used by them with Indian strangers I do not know. The first four or five miles from the Ripple River village, southwest, was across a flat bottom. In one place we passed through fine, red-topped grass, oftentimes head high, that would cut as hay from two to four tons to the acre.

Passing over a slight ridge, we came to a very warm, running stream, while, just beyond, a lake on its course was fairly steaming with vapor, from which I inferred that hot springs emanated from its bottom, as none were to be seen anywhere, and a little farther on, where we camped, the incoming creek flowed water as cold as ice. My party now consisted of four whites, eleven Indians and three dogs, a number of the latter having been left behind.

The 19th gave us an early start, and we were soon on top of a high "hog-back," from which the course of the White and Ripple Rivers could be plainly seen to their confluence and even far beyond; the big lake showing up grandly still farther on. The Indians call this large lake the Cho-ko-mun, but its interpretation I could not ascertain.[55] The White River was evidently very swollen, and out of its bank, in many places, spreading over considerable area, while nothing was left of its many islands except the trees protruding from the water.

So far from Selkirk a great many thunder-showers had been seen, and now and then one would give us a closer acquaintance, but now we began encountering longer-continued rains, without electrical exhibits, so characteristic of the Pacific slope, and which evidently we were nearing so as to feel its influence. That night we had to spread a small fly to cover the four white men, but nevertheless we all got fairly wet. We had no tent with us now, every possible thing we could abandon having been left behind at Mr. Harper's. Necessities, not comforts, determined our loads. There was one good thing about the rain, it had cleared the air of the dense smoke caused by the innumerable forest fires.

The 20th, we did not start till after two, as the rain had continued, and the dense brush furnished shower baths, wholesale and retail, at every turn. We saw a caribou soon after, far away on a ridge, and some Indians went after it, and, late that night, its carcass, a victim to my Winchester, that I had loaned to a native, was brought into our wet, miserable camp—only six miles from the previous one.

The 21st, still rainy, gave us another late start, seeing fresh moose signs soon after. Signs of gold were seen in the creeks that day. Four o'clock saw us going through a very deep mountain-pass, the ridge on the left (east) being most beautifully pinnacled, not unlike the castellated columns of the Arroyo de las Iglesias, or Valley of the Churches, in the Sierra Madre of the State of Chihuahua, Mexico. In this picturesque pass was a beautiful Alpine lake, which Doctor Hayes said was the first evidence from Selkirk yet seen of glacial action, its damming

---

55. Hayes' map shows this as Wellesley Lake, named by him for Wellesley College in Massachusetts. (Coutts 1980, p. 282). Hayes' sister was Professor of Mathematics at the college. The name remains in use today.

being caused by an old moraine. The Indians were very anxious to make another native camp that night, so it was nearly nine before we stopped after a beastly day, at a ramshackle, deserted brush camp, having made nineteen miles. This new stream was an imposing, unfordable river eighty to one hundred yards wide and flowing glacial, muddy water. After ten that night, we heard a signal shot up the river; and, after much fuss and firing all around, two Indian boys, wild as hares, came in. They said there was a large native party with three rafts and plenty of dried moose and caribou meat up the stream. Our natives spoke of these as the Copper Indians, who live in the region of that metal but not on Copper River. That day had been prolific of wasps or hornets on the trail, stinging many of the dogs and enlivening the humans occasionally, my precious self being mortally wounded twice.

The route from here, as contemplated by the Indians, was to raft down this river some unknown distance and then take to packing again, but I found them utterly loth to begin building the rafts, as they expected the other natives down with the three mentioned, which we were expected to buy. It took until ten o'clock to get the boys started back to inform the village above that we wanted to buy their rafts, and as much moose and caribou meat as they could spare. At three they came in sight and were soon alongside with the worst-looking sylvan symposiums I ever saw constructed and called rafts. Imagine about a cord of floating scrub-oak wood tied with shoe-strings and loaded with dogs and decayed deer-meat and, "excluding Indians not taxed," you have the picture. There were enough negotiations over them to have purchased the Cunard line or the Canadian Pacific railway, but perseverance and palaver conquered, and the rafts were ours, that day and part of the next being employed in repairing them.

One of the newcomers claimed that his tribe was known as the Muskrat Indians by others, and not Copper, as claimed by Jackson.

There was almost a frost, the night of the 23rd. We got away at two, southwest, on a four-mile current. In a mile we came to some slight, immaterial rapids on our left bank, that we had been warned against at the village enough times to have justified our looking out for a Niagara. I determined to call this "Raft River."[56] At frequent intervals its banks

---

56. Hayes recorded Schwatka's "Raft River" by the native name Klu-An-Tu. It is now known as the Kluane River. Aurel Krause first applied the name Kluane to the lake at its source in 1882 (Coutts 1980, pp. 149–150).

were of pure white sand, that form a great contrast to its impure, black waters. So muddy were they, that when the raft was submerged only the cross-ties could be seen and it really seemed as if we were floating down stream without anything under us. At four, we came to the main White River,[57] coming in from the south, and after an hour's run landed on the west bank abandoning the rafts. To our distraction a dozen or so Indians put in an appearance opposite us, and another long, irritating confab had to be endured, the two parties passing to and fro on rafts. That day we made, almost due west, ten miles on Raft River, four on the White, and but one packing on the land. Bear, caribou and moose signs were abundant near camp, some quite fresh. It was here that I saw a hungry Indian dog picking and eating huckleberries, and his energy must have been rewarded with a pint, at least.[58] At this camp (44) the White [Donjek] River is about a mile in width, but there are many well-wooded alluvial islands in the channel.

The scenery hereabouts is very much like the picturesque "inland passage to Alaska," but the distant snow-capped St. Elias Alps add a grandeur to an already imposing panorama.

Early the next morning, great excitement reigned in camp, caused by two caribou charging through it, followed by a crowd of shouting men firing guns and pistols and howling dogs, that looked like a lunatic asylum after a broken-loose menagerie. Nothing was secured, of course. Pow-wows between the Indians used up the forenoon; and as we started back from the river, we found a spare dog, which had overtaken us from the village, baying a wounded caribou that had been hurt in the morning *melée*. Its meat was added to our already heavy loads, and we pressed on. That afternoon we passed through a well-defined mountain pass of some length and camped near a deep stream, so warm that we all took a bath that would have been very agreeable but for the interest shown by the mosquitoes.[59]

There was a plain trail near this camp, and our Indians said it was the one made and used for many years by the Chilkats of Alaska

---

57. The river Schwatka believed to be the White was the eastern branch, now known as the Donjek River (see note #50).

58. The huckleberry mentioned was probably one of several species of *Vaccinium*. (Viereck & Little 1986, pp. 236–240).

59. This camp, No. 45, was located on the Koidern River, the native name for "water lily" recorded by Hayes (Coutts 1980, p. 150). The camp was near the present crossing of the river by the Alaska Highway, signed as Edith Creek (km. 1844).

trading with the Indians on the Tanana, to the headwaters of which it led. It is now but little used by them.

The Indians urged us to start early the 25th, as there was a large glacial stream ahead, and these are found at their minimum depth in the early forenoon; they begin to rise about ten, and by eight to nine in the evening are roaring torrents, unless the day has been a cool, cloudy one. Ten o'clock saw us at the stream, but it amounted to nothing as an obstacle, being about knee-deep. Throughout the day, off and on, we followed the supposed Chilkat trail, leaving it just before camping. Some of the snowdrifts we thought we saw on closer inspection proved to be great beds of pure white sand, of either volcanic ash or formed by disintegrated marble outcroppings. We were now getting high enough on the bald ridges to see our old friends, the ptarmigans, again; and from one high crest, just after breaking through a pretty mountain pass, we came in almost full view of the grand and glorious St. Elias Alps, the clouds just cutting off the higher peaks, and giving as fine an Alpine scene as can be found on the western hemisphere, if not the world.

This region is very "choppy," no high hill ranges of great length lying in any direction, so that a traveller can depend, I believe, on low passes for aiding him in making nearly any course. We camped at nine, after seventeen long, weary miles, and where there was very little wood except the stunted Alpine kinds. Everything had a most decided arctic appearance, as we crept slowly toward the foot-hills of the St. Elias range.

The morning of the 26th, one of the Indians put the tourist end of a Winchester cartridge through a large wolverine that was inspecting a marmot village near camp. Our sporting-bag was certainly getting very diversified, if not numerous. All we needed was a mammoth, reported by a former governor of Alaska to probably exist, and a *tébay*, or Alaskan ibex, to round it out so as to satisfy the most exacting.

That day we travelled leisurely, as the natives said there was a glacial stream well ahead that we would strike too late to cross and would have to ford early next morning. We came to it shortly after two o'clock. It was fifty to seventy-five yards wide, two to three feet deep, running ten to twelve miles an hour, with waves a foot or two in height, and roaring like a Niagara. The Indians cheerfully asserted it was much higher than usual, and that it would take two days' continuous, cool, cloudy weather to lower it, so that we could cross. Indian scouts thrown up and down the stream reported a better crossing about a mile below, where there were three wide channels, and there we went into camp to await what a morning may bring forth. By midnight the torrent was probably at its best, and every few minutes we could hear the enormous boulders, borne by the swift current, go bumping and thundering down its bed like the noise of a lilliputianal earthquake. The prospect of fording that stream was not flattering, while a gloomy rainfall added to the cheerfulness.

The morning gave us a lower river by about a foot, but with a downpour of heat that warned us it would soon be rising. The Indians were very sulky and reluctant to cross, delaying us much with all sorts of cowardly and silly excuses. We descended the north bank about a mile or so, and at five minutes past eleven o'clock the white men began crossing, Russell ahead, and by twenty minutes to twelve we stood safe and sound on the other side, our legs half frozen off in the glacial water. The Indians got a long, stout pole, and holding it horizontal, breast high, all took a firm hold of it, and as a compact, solid body crossed the stream, the pole being constantly parallel with the current. The upstream man was the largest Indian, whose pack was divided among the others, and he was really a breakwater to protect the others from the rushing current, and was occasionally carried off his feet, but the others, braced up-stream, gave him a good support. It was a "wrinkle" we afterward used ourselves to good advantage.

The Indians across, they began sulking about further and greater dangers over a glacier, etc., etc., but, on the suggestion of David and Peter, I gave them all the solace of a dollar apiece, and we moved on. The next three hours' walking was the best continuous stretch we had on the whole trip. It was across an evident glacial wash, level as a floor, beautifully carpeted with firm turf and the Indian kinnikinic,[60] and full of open parks bordered by pretty poplars and graceful hemlock trees. In one spruce grove, where the trail was very plain, there was a great number of sharpened sticks driven in the path, points up and not protruding over an inch. On inquiring as to the object, I was told that if a bear sauntered down this path—a not unusual occurrence—these sticks absorbed his attention to an extent that he showed no interest in anything else, right or left, and generally concluded to take a trot and get out of the district. On both sides of the trail, not far away, were Indian

---

60. Kinnikinnik or Bearberry (*Arctostaphylos uva-ursi* (L.) Spreng.). (Viereck & Little 1986, p. 230).

meat *caches* on high poles, that *ursus major* might have pulled down and destroyed, if he had noticed them, but which diversions saved them. I had always had a high regard for the ingenuity of the inventor who put spurs on scratching hens, so when they got to work the spurs walked them off the flower-beds, but I have transferred all this to this Indian inventor, whoever he may be.

Along the muddy shores of the many creeks cutting through this plain were gravel deposits almost wholly made up of white pumice-stone, oftentimes round as marbles and nearly as light as seasoned wood.

Jackson, as guide and *ex-officio* "medicine-man" for the party, said we must not fry grease in our pans that day, or the ice of the glaciers will tumble in as we cross and kill us all. So far their simon-pure super-stitions had not greatly interfered with us; it had rather been their acquisitiveness and laziness; so we easily catered to this and told them we would forbear oleaginous condiments rather than have a ton of ice tumble in on us.

About the middle of the afternoon we began ascending the moraine of the glacier, the exposed gravel-beds full of beautifully variegated rocks. Once on top the moraine, one hundred and fifty to two hundred feet high, it was densely brushed and covered with deep moss, with many pools of port-wine hue here and there. At five o'clock we descended from it to encounter a roaring glacial stream one hundred and fifty yards wide and three or four feet deep, that a river steamer could not have crossed. These Indians are religiously fearful of glacial ice, but it was evident to all that it must be essayed to get around this roaring river that came out from under the glacier about a half-mile above us. They besought us to make no noise while on the ice or the crevasses would open wider and swallow us up. After ascertaining there were no other solid, substantial reasons for vitiating our life-insurance policies beyond grease and noise, we retraced our tracks nearly a half-mile, and swung around over the glacier,[61] the path, as usual, over moraines resting on ice that protruded everywhere, being simply frightful. It was nearly seven o'clock before we got across the glacier, almost worn out with fatigue, for the Indians were such perfect cowards on the ice that they kept up a constant race, and never stopped a second until all were

---

61. Hayes recorded the native name, Klutlan, for this glacier. It is the source of the Generc River, and a major source of the White (Coutts 1980, p. 150).

over. They firmly resented even our whispering, so fearful were they of its consequences, while they gave all instructions in signs, and, at one point, where the party started a small avalanche down a slope of ice, they nearly stampeded with terror, whereat the scene got the better of Bowker's sense of the ridiculous, and he laughed out until all in sight turned pale; and I think they have not forgiven him yet. Before crossing, they all "made medicine," and no doubt it saved many valuable lives. Their fear of glacial ice is too pronounced and manifest to be based on any general physical reasons, and must be accounted for wholly by superstition.[62]

We camped at twenty minutes past seven o'clock on a mountain torrent too wide to cross at once, but so full of big boulders that pole bridges were easy to throw over it. *Tébay* signs were numerous near camp, but these Indians call all mountain sheep and goats, *tébay*, so that no definite animal as to goat, sheep or ibex could be identified thereby. We afterward found this true also of some of the Copper River Indians. Our natives now told us that next day we would come to the copper deposit.

The trail of the 28th. was over high, barren ridges with some timber in the valleys between. There were many ptarmigans, and signs of other game, especially moose-horns. The latter were split into bone prongs by the natives to dig copper with, so they said. Huckleberries were unusually numerous, fine and large, and I hesitate to state how many we ate, for fear that my readers will think I ought to check my estimate until I come to another fish story.

At five o'clock we went into camp on a high bank with the Copper Gulch facing us as it came out of the St. Elias range, about four miles away, and a big stream under our feet. A herd of caribou was seen opposite, and our Indians crossed in pursuit, securing three, one full-grown animal being killed by the Winchester shot-gun, while on the run, nearly one hundred yards away.

---

62. Although glacier travel can be dangerous, Schwatka's party apparently crossed without incident. The origin of the Pelly River Natives' unreasonable or superstitous fear of glacier ice and the connection with frying grease is not known. In contrast the coastal Natives from Yakutat Bay apparently had no such belief (Williams 1889, p. 393).

SIXTEENTH LETTER

At this camp, No. 49, we were sixty-two miles from White River and two hundred and twelve from Selkirk, according to Doctor Hayes' dead reckoning. The Indians were up early the morning of the 29th, and stringing out southward after copper, they said, with their moose-horn picks in their hands. Shortly after the white men followed them, and an hour or two later came to the Indians digging in the gravel wash of the west bank of a small stream that came out of the Alpine Gulch.[63] About every half-hour one of them would be rewarded with a nugget from the size of a pea to that of a walnut, and during the day, probably, they got twenty to thirty pounds among them. To those who had hoped the oft-talked-of copper mine would amount to something lucrative, the realization was quite a disappointment, but even the most enthusiastic at the start had slowly lost hope, as the true character of the Indian packers for exaggeration, and especially of Jackson, the main guide, for down-right lying, became more and more manifest. Had a mine been found at this point, there is no doubt but that it could be profitably worked, if rich enough. The Indians seemed to appreciate the disappointment, too, for they were over-anxious to return, urging a score of reasons, half of which we knew, and all of which we suspected, not to be true, to start at once; and again seemed staggered when they knew that the original party of three whites was determined to reach the Copper River to the westward, whether they went or not. They

---

63. This was Klet-san-dek, a Tanana Indian name meaning "source of copper stream" (Coutts 1980, p. 148). It was the source of copper reported by the Pelly River Natives. Schwatka named it Faloo Copper Creek because of the disappointing amount of copper found. Hayes recorded the name as Kletsan Creek, as it is still shown on recent maps. Its junction with White River is in Alaska, just west of the Alaska-Canada boundary, about 140 airline miles southwest of Fort Selkirk.

were so persistent, that Bowker assented to start with them, and at half-past six in the evening, after a hard day's walk (and work) for them, they got away on their return journey, and, with the exception of Mr. Bowker, a most genial companion, we were all glad to get thoroughly rid of them.

That evening until late was one of preparation. We divided the effects into three classes: that which we took along on our backs, that which was valuable enough to *cache* in the branches of a spruce-tree, and that which we threw away, the latter consisting mostly of provisions, and especially the great accumulation of caribou meat of the day before. We *cached* no provisions except salt, as they are a temptation to rodents to destroy the other material hidden with them, while salt is the greatest boon a lost frontiersman can find, should any ever find themselves in this locality. So far, blankets had been torn in two, field-glasses, revolvers, cameras and photographic plate-holders, sextants, artificial horizons and other scientific instruments had been left behind, while many articles I will not inventory were, or, later, had to be abandoned, so the reader can see we were down to "bed-rock," to use a miner's phrase.

The new arrangements put about seventy to eighty pounds on our backs for packing each, and the subject came up: Was it not better to double packs of fifty to sixty each, or one hundred to one hundred and twenty total, and make much slower headway at the expense of time, the element which we had the most to spare? When the packs were down to about seventy-five pounds again, by the eating of provisions, etc., we could resume single packing, and be that much further on the trail, with an equivalent start to the one from here. The bitter experience of the Takou trail prevailed, and we started single-packing at once.

The constant showers and threatening, lowering weather continued as we pulled out from Camp 49, on False Copper Creek, at half-past eight, on the morning of July 30th. Our course now kept us on the barren foot-hills of the St. Elias range, that chain, covered with drifting fog, being seldom visible, except the enormous glaciers that peeped out from under the low clouds. I could see enough, however, coupled with my experience in my 1886 expedition on the other side of these hyperborean Himalayas, to demonstrate, as would be expected, that the interior glacial system of this chain is not nearly so formidable and imposing as the sea-facing side. By keeping close to the great range, we secured better walking on timberless plateaus, but these were often cut through by impassable streams, which lower down, in the timber, spread out and were fordable. In the timber, again, the moss was so deep and soft and the brush furnished such a perpetual shower-bath on rainy days, that this selection was anything but agreeable. By lunch-time we had the cheering assurance we had made five miles, a distance we thought doubtful for a whole day, with such loads and over such a trail. Most of the morning's trail was over barren but well-turfed rolling sand (volcanic) dunes, lying piedmont-wise, so to speak; but at lunch we descended into the timbered bottom. Here we came to a recently blazed trail, but lost it off and on. Eight miles by half-past three for that day, in a miserable rain-storm, were good work, and fully equal to fifty, without loads, on a good road. Some of the trees we passed that afternoon were eighteen to twenty-four inches through, an unusual size for that country.

We sent up signal-smokes about an hour apart. There were some chances of Indians being in the country, and a packer or two to the Copper-River Indians would be infinitely acceptable, while a guide through the formidable ice-clad, snow-buried St. Elias range was very important. We had understood, from the best obtainable authority, that there was only one pass through this range for hundreds of miles on either side of our position, so the chances of missing it in this lowering weather were not improbable, even if it could be made at all in the summer, as the Indians affirmed we could not. In case we did miss it, we could cross over to the head of the Tanana, descend it a short distance, then make the head of the Kouskokqvim River and reach Bering Sea; a very formidable undertaking for so late in the season.

There was rather a singular phenomenon shown in the little creek, about a foot or two wide, near which we camped. There was a short reach of five to ten feet where the water ran backward. It was explained

[Diagram] showing the flow of water on up-stream bank at Camp 50. The arrows show the flow, the dotted part being under ground.

only by getting down into the deep bed and examining closely—the accompanying diagram aiding the explanation. The swift stream had cut under the dense, strong turf, and emerged into the old bed at such a height that enough returned by the old course to simulate an appearance that the main body was returning, while the places of entrance and emergence of the main stream were adroitly hidden to a casual observer.

The 31st was rainy throughout. The forenoon found us fording a very wide glacial river, split up into a score of channels that took over a half-hours' wading in the ice-water to cross. Then we came to a lake-like river, with ice on it four to five feet thick. I think this was a deep snowdrift in the winter, saturated by the early spring overflows, which the freezing had not yet thawed completely. Such cold, rainy weather as we saw it, it would disappear very slowly.

That afternoon, at four, we came to a glacial river fully a mile wide, and which consumed over an hour's constant hard work crossing its half-hundred channels, some of them so deep and swift that we had to take a common pole to fight our way over. Our legs were benumbed with cold, so we built a roaring fire and went into camp, with eight miles to our credit. A looming glacier could just be made out to the south, through the thick rain and fog.

The night that ushered in the month of August was one of fearful down-pouring rain. To persons wet to the skin for the last week almost, the average reader might think it would make little difference, but the traveller who has warmed up yesterday's rain held in his clothes knows that it is far better than to-day's installment of cold sleet. By three in the morning the flat ground could not carry off the water fast enough, so we were drowned out, and, arising, made a fire under a tree that gave a little protection, warmed some tea, put on our soaked packs, now weighing fifteen to twenty pound more, and started. Everything was swollen, from our shoulders, where the pack-straps cut, to the mountain streams, that were often hundreds of yards wide and flowing through the timber and brush. The dense brush itself was "wringing wet," and we were glad to get out of it and into the streams up to our middle to escape its perpetual shower-baths. Singularly enough, when we stopped for lunch, Russell had to travel an extra mile or so, wading three or four wide streams, to find water for the tea, all streams we had crossed so far being almost pure mud and sand. That evening we sent up signal-smokes again, for it seemed to be clearing a little, and I think, if an exacting

Indian or two had come along about then, they could have named their price as packers. The 2nd gave us some hopes of clearing, but the brush was yet so wet that we decided to leave it, if possible, for a gravelly river-bed to our right a mile or two, and which, reached, gave us such good walking that comments on the previous course, in not seeking it before, were in order. This river-bed was nearly two miles wide, but only half its width was now occupied by water. It gave us a good view of the country, and we could see that its bed probably determined the pass through which we must go, as it kept bending slowly into the mountains. Across this river, low, rolling hills were to be seen, and, once over them, the head of the Tanana would be attained, so our Indians had told us.

Late in the afternoon we had opened up the country ahead of us enough to see that the river came out from under two enormous and separate glaciers ahead.[64] Attempting to make the west bank, where the river-bed is about two miles wide, the swollen stream drove us back, the water being very swift and deep, and we were forced over the rough, mountainous spurs of the eastern slopes, where it was hard enough for an ibex or a mountain-goat, let alone white men with loads of half their own size. In many of the deep mountain-gullies the slopes were nearly perpendicular, of jagged rock and covered with intertwining brush. We camped early, after nine and a half miles, equal to a ten-fold amount on a good road, in order to cook three or four days' prepared rations in crossing the glacier, as we might be that long without wood. This day's weather was a little better, but the evening gave us a fierce wind off the St. Elias glaciers that cut to the bone with cold, in our yet damp condition, and that continued all that night. A few arctic grouse or ptarmigan, some marmots and a couple of plover were seen near our Alpine camp. The 3rd we started early and forded the stream debouching from the first glacier, now much lower than the afternoon before, and struck the moraine of the second glacier so as to avoid its much greater stream that had turned us back, as narrated. The mountains on both sides were serrated, pinnacled and picturesque in the extreme. The first mile on this glacier gave us comparatively good walking, about equal to our roughest hill-walking at home; but after that the pedestrianism was punishment, pure and simple, until the white ice was attained, about

---

64. The two glaciers are shown on recent maps as the Lime and Middle Fork. Neither was named by Schwatka or Hayes (Orth 1968, p. 576, 638).

noon. There were many immense, funnel-shaped holes of ice, the proximity of which we had to avoid, since an unexpected slide into one was obliteration of the most complete character. At one point we saw a complete circular arch of ice, the circle as perfect as the work of man could have been, and its radius being fully fifteen feet. The white ice of the afternoon's walk was very hummocky and rough, an unusual characteristic of this element, which generally gives the only bearable walking on these rough surfaces.

Early in the afternoon a pass in the right-hand range began opening, and by the middle we could see that it was as important as the one directly ahead, which in the morning we had thought was the only way. It seemed to have an outflowing river from under the glacier running southwestward, which, if true, was probably a branch of the Copper.[65] By four o'clock we could see it was the more important pass, with probably one chance in a dozen that the left-hand way might be the proper one to reach Nicolai's village, as marked on Lieutenant Allen's map. A still closer inspection confirmed all conjectures, when a little later we passed off the hummocky highway of ice, having made eleven miles, nine and a half of it on the glacier.[66]

The 4th gave us about our pleasantest day in the great pass, but even its eight miles were rough enough. Many glaciers put in from all sides, half the hillsides seeming to be covered with them, and at noon we lunched in front of one of unusual size, an enormous mass of ice, that came down to the very valley's level and reared its white head without any interposing moraine, the same as if it fronted the sea. I named it the Frederika glacier after my little daughter at home.[67] While eating lunch, the doctor casually remarked there was a large grizzly bear picking huckleberries on a ridge some two hundred and fifty yards away. He had not seen us, although the ground between was level and

---

65. The southern pass is shown on recent maps as Chitistone Pass. Schwatka's party continued to the westward, through Skolai Pass and down what is now known as Skolai Creek. According to Hayes (1892, p. 135) Skolai was the name by which the Copper River Chief, Nicolai, was known to the Yukon Natives (see also Orth 1968, p. 884).

66. This large glacier at the head of the White River was named Russell Glacier by Hayes to honor his friend and colleague, Israel Cook Russell (Orth 1968, p. 821).

67. Hayes 1892, p.153, noted that the Frederika Glacier was especially interesting, "in that it appears to be the only well marked case among Alaskan glaciers of active advance at the present time."

open, and, as Russell had done some extra good shooting at the moose, he was handed the Winchester with advice "to do some more." He got about halfway to him before the burly, stupid brute saw his approach, and, as he was staring him out of countenance, the Winchester spoke. During the day while travelling, or in some unknown way, the rear sight had been elevated, and the bullet whistled over the bear, but so close, that he did not wait for a second, and, wheeling, ran along the glacier's front and started up the great mountain slope until lost in the high fog; his speed, at first, being greatly accelerated by two or three more shots Russell sent after him.

After lunch we essayed the glacial river emanating from the Frederika glacier, and got across the first two channels, but the third and main one was plainly too dangerous to attempt. Every minute, or so, a great ice-jam would break loose in the glacier and come floating out the deep channel, the blocks from a foot to five feet in diameter, pouring and thundering down the swift stream in a way that would easily break every bone in one's body if caught in the water during such a rush. We retraced our steps to the grizzly-bear lunch camp to await a lowering next morning. All that afternoon and that night we could hear the thundering of mighty avalanches, that shook the ground like so many small earthquakes. We were now in the very heart of the heaviest range of our continent.

We now added wolf tracks (they were perfectly enormous fellows) to our game signs, but we never saw any of the animals.

The 5th was rainy again, and although the glacial rivers had fallen some, we could not cross the one we had essayed the day before, so we returned again to the grizzly-bear camp, and crossed the main river to the south side. It was much deeper and wider but not near so swift nor so jammed with floating ice; still it was just all we could do to get over. Then our forenoon's fight over the moraine of a glacier would exhaust all the superlatives of the English language if properly described, so I will not attempt it. Once over we came to a little spruce timber, which, though scattering, seemed like a Godsend in this desolate district. Here we saw a very old Indian sign (a tree tacked down with faint sign of an ancient trail), probably made ten or fifteen years before, but it augured well, that we were on some line they travelled at least. That afternoon gave us a moraine under foot again, and more rain overhead again. This country seemed alive with bears, the signs were so numerous; but the dismal dreary weather gave us little show of seeing any kind of game. Signal smoke sent up as soon as we got to timber again may have prevented our seeing some.

We thought we had encountered every conceivable obstruction known to Alpine climbing during that summer, but we were now treated to a combination of huge, rounded boulders, overgrown with twisted black alder-brush, on the steep mountain slopes, that took the first prize; while we took for the river bottom, intending to wade down it or drown. We got hardly three-fourths of a mile on a gravel-bar before the roaring river forced us to the banks again; but we were very thankful for that, as the great Alps were now pinching into the mountain torrent extremely tight, and we were fearful that its little valley, rough and rugged as it was, might be obliterated at any time by the dizzy precipices, often twenty five hundred to five thousand feet in height, looming over us and dangerously near, meeting in a mighty canyon.[68] Every yard counted now in a struggle that a few miles would probably terminate in our being forced back or opening a more agreeable aspect.

That day Russell said he saw coneys in the rocks on the mountain slope. We camped on a large, long lake in the course of the river, caused by a glacier from the other (north) side damming it up.[69] It was fairly stocked with icebergs, most of them being at the up-stream end, showing, I think, a steady wind up the pass at this season, and which we had always found to exist. That day's nine miles were one of the memorable journeys of the trip, a fearful struggle with icy torrents, slippery rocks, hummocky glaciers, muddy moraines, willow and alder *chaparrals*, soft mountain turf knee deep, dizzy heights on rock slides, and in every way one long to be remembered.

After the evening meal, the doctor walked ahead to the high crest seen from camp, and on returning reported the country apparently opening and, seemingly, no great obstruction ahead.

The 6th was another hard struggle with Alpine difficulties. We struck a well-defined trail early, but whether game or Indian, was hard to tell, probably both, with a great preponderance toward the animals, as judged by the relative signs. Mountain goats were seen on the high southern

---

68. Previously cañon.

69. The lake mentioned by Schwatka was evidently the result of damming by Nizina Glacier (see Hayes 1892, p. 154). It has since disappeared owing to retreat of the glacier and is not shown on recent maps (see Orth 1968, p. 692).

cliffs, while we were at the mid-day meal. An animal with palatable meat would have been very acceptable just then, but this goat meat was a little too high for us. When we came to the end of the glacier front, or rather to the end of the moraine, it had pushed completely across the river, and where the roaring river came out from under it, we *debouched* on a beach that gave us over three miles of comparatively fine walking, and made our day's record twelve miles, a most gratifying result under the circumstances. From this camp (56) the prospect ahead was dubious. Enormous precipices flanked the winding stream that cut back and forth in the cañon-like valley from one wall to another, forcing us to cross it every mile or two, or play mountain goat over the stupendous spurs. After emerging from under the glacier, the great river rapidly spread to nearly a mile in width, but abated not a jot in its swiftness. A spruce grouse and a hatfull of vinegar currants added somewhat to our sparse ration supply.

The weather was now getting better as we cleared the principal mountain range. My diary of the 7th starts with "I look on this as the determining day as to whether we shall get out soon." We got away early, but after a few hundred yards that brought us out of the spruce woods in which we had camped, we saw that the river concentrated into one swift, surging channel, cut into a perpendicular cliff one thousand five hundred to two thousand feet high on our side, about a mile ahead. As the river where we stood was spread out into a great many channels, probably a mile in width—and a foot lower than the night before—we started across at twenty-five minutes to nine for the west bank, and reached it at ten o'clock after continuous work, being compelled to turn back from a number of the worst channels, and often obliged to use the big camera tripod for a pole for the whole party, so as to fight the swift current as a compact body. As we crossed this wide, open space we could see high snow-clad mountains far ahead of us, a most discouraging sight. Their great distance was the most encouraging part, as it was known that the upper Copper River tributaries were inclosed in a great interior basin, and that the main stream broke through this rim of mountains not far from the Pacific coast.

Glacial rivers flowing mud were now coming in from both sides, but the glaciers themselves were farther and farther back in the side valleys as we descended, whereat we heartily congratulated ourselves that their attendant moraines and other difficulties were probably being left behind forever.

After crossing, the river forced us from the gravel beaches back into the timber where a combination of its intricate underbrush and the boulders that had been thrown down from the high slopes presented the usual obstructions, but as they were on the level they were at least passable at the rate of half a mile an hour with severe labor. Early in the afternoon, however, we came to the straw that broke the camel's back, and that forced a new vision into the history of the expedition. A great bluff before us overhung the river for some two or three miles, but we thought we could make along its steep wooded talus to the apparently open country beyond. I will not use the necessary strong superlative adjectives to properly describe that two miles of misery, but will condense its many hours of mortal agony into a minute's description by saying that it took us until dark to get along it, some six to eight hours. It was so absolutely necessary to reach level land before dark, or roost in the trees with the birds, that not over a half-dozen rests of a minute or two each were taken on this trip, mere gasping spells when utterly breathless. In one place the moss carpeting, growing over a prostrate log, leaning along the steep slope, gave way limbs and all, under my heavy weight, and I disappeared from sight, my pack catching on a huge limb that alone stopped me. My feet dangled among the shale and shingle gravel below that, thus stirred up, started to slide down as a small avalanche which, accumulating, threatened to toboggan the log, the explorer, the moss and the mountain into the surging rapids below. I objected to the second item being lost; so, after some athletic gymnastics, I managed to crawl out, having lost the ax off my pack, which I had to regain at a peril that I would not have risked for a whole ax-factory under any other circumstances.

That night's camp looked gloomy; the rain had been falling all afternoon, and most of the party had been doing the same thing along the slope, until, tired, tattered, torn, dirty and dejected, and in some cases bloody, we looked worse than thieving tramps that had escaped through hedge-fences the dogs had gone around. We had carried pack-mule loads for eighty-six miles through the only pass in the roughest range of North America, and our misery was getting monotonous and our whole natures cried out for a change of programme. We got it.

SEVENTEENTH LETTER

On the morning of August 8th, at Camp 58, we came to the conclusion to build a light canvas boat and descend the river.[70] The prospect was far from flattering, but that prospect had to be compared with the one of packing with almost shoeless feet through the most infernal country that a man ever struggled, at a rate of speed which would have shamed even a messenger-boy. We were already practically without provisions, and game was scarce and harder yet to get.

At this point on the river, the current was very rapid and turbulent, waves running one to two feet high in the swiftest parts of the channels; but there was a noticeable improvement over the violent torrent seen near the glacier, some five or six miles back, and if this ratio continued, as we hoped and believed, another equal distance ought, at least, to let us out of dangerous water, if not into a tranquil stream, and then Indians should be met in a few days. In the way of material, we had a couple of pieces of light (eight-ounce I believe) canvas, that the doctor and I had used as water-proof coverings for our packs. It was dangerously frail for such a trip, but the best, and in fact, the only material we had. For thread to sew, we used mostly the fish-lines we carried in our pockets, the small amount of pack-thread we had for every-day repairs being inadequate. Our tools comprised a couple of sail-needles, hand palm, a pocket-kit of tools and a camp ax, besides such hunting-knives and pocket utensils as any field-party would usually have. The framework was to be of wood; the keel, bow and stern of spruce and willow; and the ribs and gunwales of black alder. All of this was green and

---

70. Camp 58 was on the Nizina River a few miles above the mouth of the Chitistone River.

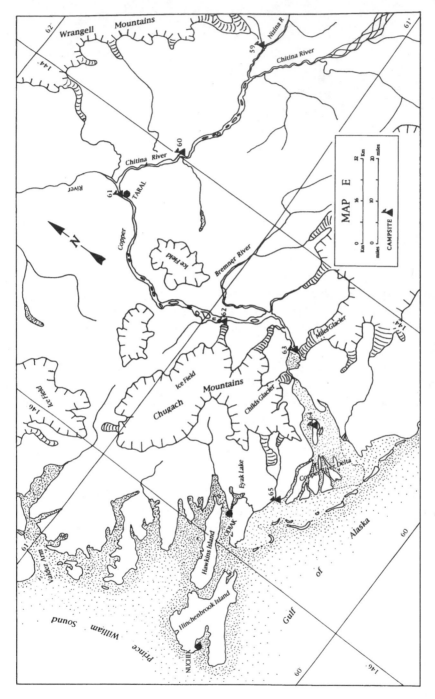

Map E: Route of the New York Ledger Expedition of 1891 from Camp 59 (August 10) to Nuchek, Alaska, (August 24). Topography based on U.S Geological Survey Alaska Topographic Series: Bering Glacier, Cordova, McCarthy, Valdez.

heavy, but in our favor. Russell was a practical woodworker, and had often constructed boats for frontier purposes, under all the difficulties that usually attended such construction, while the doctor was quite handy in "all-round tinkering." The boat was made fourteen feet long and nearly five in beam, giving it a wherry-like appearance, a very good form for the choppy waves of the rapids. A little bit of lard had been religiously saved for just such a contingency, and this we mixed with spruce gum collected from the trees, to water-proof the bottom of our new craft. By the night of the 8th, Russell and the doctor had made the frame of the boat, while I wandered in the wet woods and gathered a gallon or two of gum, and, as a result of my picnic, got an extra ration over the others of a pint or so of rose-buds.

In a swamp near camp I saw "devil-sticks,"[71] a sure indication of the moist Pacific coast region, an indication which an all-day's rain confirmed. My diary for that day closes with: "We are nearly down to rose-buds and raw flour, but our new boat will now swamp or save us in a short time."

By the night of the 9th, the boat was finished, but when the pitch covering was daubed on, it was clear there was not enough to cover it properly, and the early morning of the next day was used in collecting more for this purpose. The time was hardly lost, however, as we intended to wait until the late forenoon for the morning's rise of the fluctuating glacial river. By ten the boat was done and laid on the bank, while we began packing up. We could now take in the boat and the big waves beyond at the same glance, and it certainly looked like sending a boy to do a man's work, but when one has only a boy to call on, the little chap may "beat nothing all to glory," as the Arkansas backwoodsman says.

Except in foot-gear to protect us on the sharp gravel, if we jumped out, we practically stripped to our under-clothing, to meet any accident, all valuables—and we were about reduced to where everything was valuable—being tied to the cross-pieces so in case of upsetting they would not be scattered over too much unknown country. We got away at half-past ten, but used a half-hour maneuvering around gravel bars

---

71. "Devil-sticks," or Devil's club (*Oplopanax horridus* (Sm.) Miq.) is a large plant noted for its spiny stems and large spiny leaves. It is common in dense thickets on moist soils in woods along the coast of southern Alaska (Viereck & Little 1906, pp. 197–198).

so as to get well out into the central channels and clear a rugged point some three hundred yards below us that we had been seriously discussing for the last two days as sufficient to wreck an iron-clad skiff. Here the water boiled over jagged rocks and intertwining drift-timber in a shallow sheet of foam that we heard roaring like a cataract during our stay at camp. Had we known our full capacity for such obstacles, as determined by the next few days' experience, we would probably not have lost this half-hour but put the canvas-back right through and taken a nap while on the way.

At eleven we were all ready to try our tissue-paper transportation. Everyone jumped in, the boat leaping like an unbroken horse held on both sides till mounted, and we started forward at a gait of from ten to twelve miles an hour, the islands fairly whirling by us in a jumbled mass of gravel, boulders and drift-timber. It was dangerous to attempt to stop so we cut through channels and dashed through breakers that were better than barbers' elixir for raising hair. Russell tried to stand up in the stern to see what harbor in Hades we were heading for and an unusual lurch of the plunging boat sent him spinning like a glass-ball from a trap, and there was considerable activity on part of "all hands" to get back to our normal status, plus a good ducking. Inside of a few minutes both Russell and the doctor were pitched out trying to stop the boat to inspect a very bad outlook ahead, and all alone in the world I started out at a speed that promised to land me in San Francisco that evening if nothing interfered. All the poles had been broken by this time, and the two paddles were lost by the same accident, so I considered myself lucky when I pulled up on a gravel bar not far below and awaited the two mariners who had to lock arms to cross the swift channels between us.

We were certainly fortunate in this last accident, for both the channels outside of either party were roaring torrents, that Leander could not have crossed. Had we taken either of them, however, we probably would not have struck bottom and so nearly upset. It was just noon when we ripped the first hole in the bottom, turned sidewise in a twinkle, dumped Russell overboard and came near rolling over and over down the steep chute as a wrecked expedition. We were nearly an hour repairing damages. At half-past two, we managed to land on a small bar with some dry driftwood, and here we ate our lunch; our whole course, so far, about twenty miles, in a little over two hours, being a repetition of the first hour I described, having been in and out of the boat a dozen times either to inspect threatening outlooks or when thrown out by the plunging craft itself. How the balloon-like boat stood the banging and severe straining was the greatest mystery of all. It had been used roughly enough to have wrecked a canal barge, it certainly seemed.

The aspect ahead was now very black and threatening; the great river, of over a mile in width in places, was there narrowing into a colossal canyon of one-tenth that width; the rough roar of the resounding walls being carried to us on the breeze and sounding dubiously like a cataract or whirlpool torrent. Russell got over to the north bank and ascended its precipitous sides, spending nearly two hours ahead inspecting the canyon; but he could find no fall of great size though he reported the inclosed roaring rapids as dangerous enough to rank with one, unless we were very lucky in our management through it. While he was gone, signal-smokes were sent up to attract the attention of any Indians idiots enough to find this country congenial.

It was a hydraulic hurdle-race for the next five miles, and I do not believe we could tell, half the time, whether we were under the water, on top of it or flying through the spray. The wide beam of the boat made it impossible for me to paddle from my place in the center without adding danger to a peril already skirting around the edge of destruction, while all this fell to the others fore and aft. I can do as much hard labor to avoid work as any one, but I was never placed in a more trying position in my life than in such a place where probable disaster or even death pivoted on muscle, and yet could not turn a hand to avoid an accident that might mean wiping out the expedition, while the others were often nearly breathless and powerless with over-exertion. No better relief could have been devised, in such excitement, than equally exciting labor.

By half-past four, we had to make another inspection, the canyon narrowing to half its original size and towering to double its former height. It was so contracted directly ahead of us, so black and shadowy, and turned such a sharp angle, that it seemed as if the whole frothing mass plunged into a tunnel or subterranean abyss and disappeared at once. We were not yet entirely out of the glacial district, and the thought of suddenly turning a corner and disappearing under a glacier impinging against the stream hung over us as a mournful addition to the more tangible dangers.

Just after five o'clock we tore another great hole in our canvas craft, and came near sinking before making a bar where it could be repaired, they were so very scarce in the narrow canyon. By the time it was done, it was decided to make a long stay here, as the sun was now in our faces, on the general course westward, and we had to add its dazzling glimmer from below and pyrotechnic rainbows in the spray above to the already long list of confusing and exasperating impediments.

We got away at seven in exceedingly heavy water, and at twenty minutes past seven, seeing the canyon ahead apparently obliterated in the black shadows, we stopped on a little bar, having made a half-hundred sharp, angular reaches in the short run. As we found a little wood on one end, I decided to camp here after our very exciting day, with thirty-six miles in our favor in a less number of hours than it would have taken days to pack it, but with an inverse ratio of danger. My diary closes: "A day of dangers long to be remembered, being a continuous run of rough rapids, with waves often four and five feet high and tossing in all directions."

Our barometers now showed that we were but seven hundred feet above the sea-level, a drop of some two thousand five hundred to three thousand feet from the interior plateau of the White River basin.

Despite our descent from Alpine heights and the midsummer season, we had a tinge of frost the night of August 10th-11th, and the morning broke clear and cold. We now had but one day's full ration of flour left, after which we could help ourselves to the salt.

It was twenty-five minutes after ten before we got started the morning of the 11th, but from the earliest light we had been busy. It took us until nine getting across the stream by tracking, wading and other maneuvers, so as to inspect the unusual aspect ahead. Russell's report was the worst one yet received. There were five to six miles at least of much heavier rapids and more formidable canyon than any encountered so far, and his general opinion as to the situation was summed up in that "there is not one chance in a hundred for any sort of a boat to get through safely." The roaring rapids just ahead, probably equal to the whirlpool rapids below Niagara, was but one of a score or more in the course, and the original plan that I should get out, to lighten the boat, and make my way over the northern mountain-spur and trust to rejoining lower, was given up as impracticable for that reason. I believe we all thought later it was well we did so, for in the rush through the far-reaching rapids my weight, as rapidly adjustable ballast, was probably one of the determining causes of safety, especially when mounting the high curling crests of the breakers, where a lighter craft, or even one with equivalent dead ballast, would have been whirled over and around like a shingle in a cyclone. It may have been fun for the paddling-boys, but it was worse than death to the ranine ballast. The high canyon walls were perpendicular now, as an average, and quite as often were projecting over and above us, as they slanted backward where a mountain-goat or a desperate explorer could ascend. The course of the canyon was quite a tortuous one, and, although the reaches were short and the turning angles sharp, yet they were somewhat uniform in their proportions, and a little experience soon taught the paddle-men how to manage them to the best advantage and to the least danger.

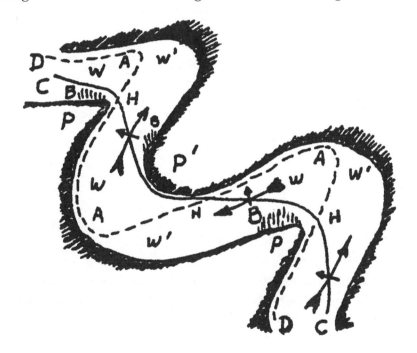

Probably the above diagram will explain the idea better, the arrows indicating the direction of the water's flow. The broken line DD would show the line of maximum danger, so to speak, and this danger was greatest again at the acute angles A, A, A, than at any other points. Here the swift glacial water, a perfect sand-blast for centuries, had often cut far under the hard rock-walls against which it impinged, sometimes at right angles, until the yawning cavern underneath looked like certain death itself, as we shot toward its entrance covered with dashing spray, that generally hid it until within a paddle's reach. These impinging points were always flanked by huge and dangerous cyclonic whirlpools, often spinning so like a buzz-saw that the vortex was two to four feet below the rim, the outer whirlpools, so to speak, or W', W', W', being generally more perilous than W, W, W, or the whirlpools inclosed by the acute angle of the danger line. Where the rocky points P, P, were acute, the danger was greater beyond, as would be expected, than at P', P', where they were obtuse. The highest waves, often four to six feet, were just off the points at H, H, H. While seemingly so to the novice, they were not the most dangerous, however. B, B, B, just around the points, show where bars were occasionally found, but seldom so frequently as I have marked. These were our life-preservers in case of upsetting, ripping holes in the boat or any other accident. The course of the canvas boat is roughly shown by the line CC. It will be seen that this took us through the highest waves, which we at first avoided, but nearest the safety-bars, and really escaped all the most perilous parts. No paddling was done and none was needed to gain direction in the lightning-express current, but all energy, and it was terrific, was bent in making from point to point, as shown by the line; so we really ferried backward and forward, so as to gain P, P', P, P', while the current carried us along. The boat's bow was, therefore, always pointing toward the shore or canyon wall which terminated in the capes P, P', etc., we were endeavoring to make, as shown by the featherless, stubby arrows. This made the boat turn completely around at each bar. But in going through the high waves, at H, H, H, this perpendicular position to the current was perilous in the extreme, as the curling breakers could easily have upset a better-made craft of four of five times our size, so we really began the turning around just as we entered them. It must not be inferred by any means that there were no other waves elsewhere in the canyon; it was simply a boiling caldron colliding with a heavy ocean surf the whole distance from end to end, and from wall to wall. My diagram and description are but a general idea to which we were meeting exceptions every few minutes in the perilous passage.

At many of the bars, the boat was stopped to rest the fagged paddlers or to inspect some particularly ominous outlook. At twenty minutes past eleven, we came to a place, an unusually rough, jagged point and sharp turn, that nearly frightened away the little wits we had left. There was a clean, sheer fall of five or six feet on the outward sweeping curve that faded to a foaming rapid at the salient point of the wall opposite. So far the boat had been acting splendidly, a perfect duck in a deluge, but to shoot falls as well as rapids, was asking entirely too much of it. The rapid on our left curled around to the foot of the fall on our right in a stupendous whirlpool, greater than any yet seen; a perfect cone of swirling water, twenty-five to fifty feet across, that grasping a great log, a dozen to a score of times our weight, spun it around like a straw, and sucked it out of sight in the vortex as if had been one of lead. We were not over twenty yards from the fall when it was seen, and we all expected to go over it, but unusual exertions put the craft to the bar opposite, where we got out and let the boat drop around by painters fore and aft. We were now feeling the pleasant reaction of confidence in our frail craft, and as after the fall was passed the canyon began opening a little, we took advantage of the first gravel bar where wood was to be had, and took a long rest and a short lunch, our provisions being our most menacing embarrassment now. Our first pleasant surprise in the canvas boat was in an unusually swift current that carried us against the canyon-wall opposite, where we felt sure we were doomed to an upsetting at least, or more likely to an utter collapse of the craft. The exertions of the paddlers seemed to have no effect to retard the speed, and we went broadside on, but the high, curling water that the wall dashed back was a perfect hydraulic buffer, and as easy as if done by giants' hands, the boat was stopped and thrown back into the waves. After two or three such experiences, we felt less and less reluctant to dash up against these seemingly dangerous places. Undoubtedly the escape at the whirlpool fall was our closest, that day, and certainly narrow enough to satisfy even a dime-novel hero. Notes and observations were almost completely out of the question, as the time could not be spared, and the dancing, plunging, whirling boat rendered it impossible. We jotted down at the bars, where we rested or lunched, what we remembered from the last stopping-point. At the lunch-bar (not a railroad one, though like it as to provisions), I find jotted down in my journal: "A perfectly fearful forenoon, with enough danger to turn a hair mattress gray."

From this point, seven miles from our morning camp, we could see that our vicious jagged canyon opened into a much wider and more pretentious valley, about a mile ahead, evidently carrying the main stream into which ours emptied. We got away at twenty minutes after one and reached the new stream in a few minutes. It was evidently the Chittyná of Lieutenant Allen's map. It was a much larger valley but the river flowed less, although clearer, water. Everything seemed to indicate that we had shot a tributary during a high freshet.[72] The Chittyná was running quite high waves, but they were very long and mostly destitute of surflike caps, and in general not dangerous; the current also dropped to about six or seven miles an hour. At thirty minutes after three, we passed the mouth of the Tébay Creek, south, and an hour later Dora Creek, north; both of Allen's map.[73] A strong cold wind now came up from the west, square in our faces, and in our wet condition was very disagreeable. It was nearly six when we stopped on the north bank, and where the doctor and Russell took a look at a supposed canyon-like contraction of the river, but which amounted to nothing, however. As the wind had increased to a furious gale, it was decided to camp on the slope, although a bad spot for that purpose. Our food was nearly out, and an additional stew of rosebuds and green cranberries made Russell very sick. My ankles were stiff, sore and swollen out of all shape, owing to the constant immersion in ice-water for the last week or ten days.

We had made thirty-five miles that day, and our original five miles of rapids had drawn out into nearly fifty, which, could we have viewed them ahead, all concurred in the statement that we would rather have packed around them or starved to death, than attempt what we had done. It was a lucky thing, indeed, that we could not thus have viewed them. To Russell belongs the main credit of the boat and what it accomplished, with Doctor Hayes a close second, while I reserve only the smaller part of which I have already spoken. I have never seen a better river-man than Russell, and very few his equal in a very wide frontier experience.

The night of the 11th the river rose over two feet, and the little canvas boat that we had pulled up on the shore to dry was afloat in the morning swinging by its painter. We got away at twenty-five minutes

---

72. The Nizina River, which Schwatka's party descended, was shown as the "North fork" of the Chittyna River on Lieutenant Allen's map. The Native name Nizzina was first reported by Hayes (Hayes 1892, p. 124) (Orth 1968, p. 692).

73. Dora Creek was named by Allen for his future wife, Miss Dora Johnson, of Chicago. The name was probably first applied to the stream a few miles east, now known as the Gilahina River (Orth 1968, pp. 281, 366).

after six, and it was a grateful sight to see low wooded points protruding into the river instead of frowning canyon-walls. I notice that I yet record in my journal: "The hardest part of writing in my journal is the constant dancing of the boat in the waves of the rapids." At half-past eight, we passed through a contracted rocky gate in the river that is quite picturesque.

At nine o'clock, we came to the junction of the Chittyná and Copper Rivers, but saw nothing of the Indian village of Taral, that Allen marks at the confluence on south bank of Chittyná, but after descending the Copper some three miles, at ten minutes to ten, we sighted it about a mile ahead, and a little after ten made it, and, to our satisfaction, found that it was well peopled. Nicolai, the chief, was present and he gave us a hearty welcome and a still heartier dinner of tea, some dry Graham bread and a great slab of fried bacon each. It does not seem so wondrously palatable just now, looking back at it, as it did then; but if I could combine its palatability of that time with its present price in the great cities, I would not exchange the combination for the richest mine in the world. He sold us all the bacon he had, a small piece, and some flour, while another Indian gave us a salmon; altogether we were faring unusually well. I was greatly surprised to find that Nicolai spoke Chinook well enough to get along with him. This jargon I had believed, from my former experience on the Pacific coast, did not extend above southeastern Alaska.[74]

Nicolai told us he expected to assemble a party of Indian fur-traders in a week to ten days, and descend the Copper River to the trading stations near its mouth, and that we could join the party. We now saw plainly that if we had packed out, we would have been too late to take advantage of this, the last chance of the year for native help to assist us to the coast. A little discussion disclosed that he was willing to start a little earlier if he could get his scattered party together, and runners were sent out for that purpose. The river was entirely too high, he claimed, to risk it now, as there were many dangerous rapids ahead in the canyons that would be greatly improved by a lower stage of water. As Allen's map showed a number of canyons ahead, it was not hard to

---

74. Chinook jargon, a language of unknown origin, was used in trade on the northwest coast for centuries. It combined words from several Native languages of the northwest as well as English and Canadian-French. It gradually disappeared as English became dominant. In Alaska and the Pacific northwest a few words are still used occasionally (see Thomas, 1970).

believe his story after our recent experience. We camped near-by to the village, on a pretty, grassy bank, the doctor and Russell preferring to sleep out-of-doors than in an Indian house, with its Ylang-Ylang and Stephanotis air. They cocked up the overturned canvas boat and used it as a tent, while I took up quarters with Nicolai and "chinned" him on Chinook till far in the night, when we dropped to sleep, and I dreamed of going over Niagara in a single-scull shell and shooting the Lachine rapids without repairing the holes.

EIGHTEENTH LETTER

Nicolai and the other Indians around Taral knew thoroughly the country that we had just passed over and informed us that the river on which we built our canvas boat was called by them the Nee-zee-náh [Nizina]. They would not believe we had shot the Nee-Zee-Náh with the frail craft they inspected, and no assertions would convince them of it, and, in fact, I doubt if they believe it yet.

Taral was a town of but two or three houses, Nicolai's being the most imposing; yet there was a number of others in sight across the river and up and down the stream, while the peculiar *caches* for dried salmon, or small houses perched on poles to protect their contents from wild beasts, looked singularly like Indian cabins, at a distance.

I bought a couple of bales of dried salmon (there are from twenty to thirty fish in a bale) for the mess, while we had fresh salmon every day from the river. In fact we lived on fish essentially until we thought we had struck the Devonian period.

We had been cramped so in the boat for the last few days that we needed exercise rather than rest, but my feet and ankles were so swollen, as the effect of the ice-water poultices of the last two weeks, the black-and-blue skin seemingly ready to burst, that I was not averse to laziness, since I had so good an excuse.

My first night in a house I did not sleep well. When one gets used to making down the earth for a bed, pulling the sky over oneself for a blanket and going to sleep, even the Coliseum at Rome seems small, but an Indian hut, hermetically sealed, about six by eight, and only five feet high, with twelve children, assorted sizes, and a reserve corps of dogs, can rival a sardine-box in compactness.

The river fell during the night, but was slowly rising again through the day; the effect of the summer glacial tides as they might be appropriately called.

We got two or three little "dabs" of dirty flour from one or two Indians and then fell back on the ubiquitous, everlasting, non-failing, perpetual salmon. The Indians were now only catching such salmon as they needed from day to day, the big, or main, "run" being over, as shown by their *caches* being full of dried salmon.

The morning of the 14th of August, a number of Indians came down the river—"Stick siwashes," as Nicolai called them in Chinook. They were, I understood, a portion of the party to go with us to the mouth of the river.[75]

During nearly all our stay at Taral, the wind was blowing more or less severely, sometimes amounting to a gale, and seldom perfectly quiet. It was almost invariably from some southern quarter—usually southeast—and the Indians told me that this was reversed in the winter, when the bitter cold winds off the Wrangell Range blew even more fiercely down the river.

It cleared that forenoon, and several snow-clad peaks could be seen in this range to the northward and northeastward. One of them seemed to be smoking at the apex like a volcano. This was undoubtedly Mt. Wrangell, which was named by and reported to be a volcano by the old Russian explorers of this region before it fell to us. Several days' observations, I think, clearly established that it was yet smoking, though hardly active. It is a flat dome, and I should estimate not over ten thousand to twelve thousand feet in height. Nicolai said it was always smoking, and the aspects in the evening were very convincing.

About nine o'clock I started to retire in Nicolai's. I had to go through a sort of vestibule or outer building two or three times the size of the bedroom, and here were a dozen savages, stark naked, perched on the wide bench running around the room, all taking a sweat-bath and singing a wild chant at the very top of their voices. The room was insufferably hot and moist, the stench mephitic, and the whole combination diabolical in the extreme. Nicolai quietly told me that night of

---

75. In Chinook, "Stick" = a stick or pole; (here in reference to the small size of timber in the Alaskan interior as compared with the coast); "Siwash" = Indian or Native. Hence, "Stick Siwash" = a native of the interior (see Thomas 1970, pp. 95, 97).

the killing (by lynching, I heard afterward) of one of the Atnas as this band is called, on the head of Cook's Inlet, by some white men that spring, shortly before our arrival. I understood him to say this was the reason he was so anxious to have me, as chief, sleep in his room, as the Atnas had not yet determined fully what to do in the matter, but he thought the least would be to withdraw all trading from that district for the future and transfer it to the traders at the mouth of the Copper, where they were now going. He also assured me they would get away in four *moosums* (sleeps), unless the river got much worse, which was not likely.

The morning of the 15th, the river was very low, bars projecting as islands in every direction, and the Indians told us they might get away next day, but did not seem to make any preparations.

All my contact with these Indians showed them to be a most merry-hearted and jovial lot of people, overdoing it, in fact, in many ways. One time all of them were speculating as to whether a white spot in the high cliffs was a mountain-goat or bank of snow. All opinions had been expressed and even I was called on, when I declared it a salmon and the silly joke had the desired effect of making them fairly explode with laughter. If ever I got tired and worn out on anything in my life, however, it was on my salmon joke for during the next week, night or day, so long as they remained with us, they would look at me, blurt out "Salmon!" and the whole cavalcade would go into humorous hysterics. Both as food and fun I got heartily sick of salmon.

An Indian armed with an old Russian shot-gun killed a number of rabbits near Taral and pressed two on us, stoutly refusing any compensation. It certainly was a very non-Indian trait, or my third of a century among them has been wasted.

When night-time fell with a gale blowing, we understood that our starting next day would depend on the state of the wind, and for a wonder there was very little in the morning.

A rain threatened, however, but we got away at seven, our first act being to cross the river to land a passenger, a long-haired poetical-looking cut-throat who was returning on foot to a village far up the main stream. Once rid of the poet, we started down the four-mile-an-hour stream, the Indians, eleven in number, doing hard paddling and loudly singing an accompaniment that was not unpleasant. Instead of the usual monotonous "yi-yi-yi-yis" of most Indian chants, there really

seemed to be an air involved in the effort, that needed all the notes in
the gamut to express. The different airs in the various songs were also
quite noticeable to even my uncultivated ear.

The boat was one of a strong wooden frame, whose carpentering
was well done, and this was neatly covered with *lurlak*, or the tanned
skins of the large seal, procured from the coast. It was very similar to
the *oomien* of the Eskimo; and as the mouth of the Copper River is said
to mark the limit of the Eskimo on the western coast of America, there
may be some closer relation between the two than this apparent simi-
larity. It was from twenty-five to thirty feet in length, and, with the
eleven Indians, three whites and a load of furs and other material that
reached the gunwales, it drew a foot to fourteen inches of water and
stood about the same distance above. The first four of five miles was
through a canyon, as mapped, but it was not a formidable one at all.
Occasionally the rocky points gave a little rough water, with waves
about a foot high, and the Indians were as afraid of them as if they had
had the hydrophobia. No doubt, the long, stiff, flat-bottomed craft, so
heavily laden, was not a life-boat in rough water, but the Lilliputianal
dangers which were so energetically avoided by herculean efforts clearly
explained why they doubted our shooting of the Nee-zee-náh [Nizina]
in the little *Eli*, and had so strenuously opposed our taking it through
further dangers. Anything we met on the Copper would have been
child's play for the canvas craft we left at Taral.

About nine or ten, we could see we were getting out of the interior
basin and cutting through the rim range of mountains that separated it
from the coast. The country became quite Alpine, the hills bare of
timber and covered with intensely green moss or grass. We saw a number of mountain-goats on the high cliffs, some two thousand to three
thousand feet above us.

At half-past one, having stopped an hour for lunch, we came to a
beaver-house near the shore. While a number of the Indians got on it,
to tear it up and stamp on it to drive out the beavers, some in the boat
watched with guns for any to pop up their heads, but they were
unrewarded.

A few minutes after, we passed a large plateau glacier on the west
bank, extending off in the haze as far as the eye could reach, and was,
I understood, the one used occasionally to portage over and reach Port
Valdes, in the Prince William Sound. The middle of the afternoon we

passed a large river coming in from the east, and not mapped, as we could make out. Here we stuck on a number of mud-flats. Swans were seen, but none secured. We were now in the heart of the glacial country again, and there were scores of them, big, little and medium-sized, in all directions.

At four o'clock they stopped to shoot a mountain-goat on the mountain-tops, and were so long about it, it was decided to camp here, although a very disagreeable place among the dense willows on a steep slope. I vented my overwrought feelings as follows, in my diary: "The only thing an Indian can be depended upon to do thoroughly in this world is to pick out an exceptionally bad camp, when the selection is theirs." *Linate*, or wild celery, was plentiful around camp, and also a few salmon berries, and other indications of the Pacific coast.[76] The river here, about fifty miles from Taral, was about a mile wide and apparently quite shallow.

My diary again: "There is a fearfully penetrating smell of salmon about this camp, that no one direct from civilization could possibly stand. It ranges all the way from the future anterior of fairly fishy to the past preterit of putrefaction." A few avalanches were heard during the day, one at noon being very heavy.

Among our party was an old Tanana chief whose abode was on the headwaters of the Copper. He had a very few of his tribe, two or three, and he and Nicolai were sort of royal personages among the others. Seeing my scanty bedding, he loaned me a fine, large bear-skin to sleep on, and the other whites were also looked after. For the first night in many, we found a mosquito bar necessary for comfort.

We got away at five o'clock, the 17th, but some mountain-goats delayed us considerably. At several rocky points the Indians cached dried salmon in the crevices for their return trip. The river soon widened to a sluggish lake until nine o'clock, when we passed through a very narrow "cut-off" channel between a rocky island and a glacier.

At Taral, we had understood from the Indians, there would be a great deal of trouble getting around or over the two great glaciers near the Copper's mouth, but just how was not made quite clear, but in some way a rapid around or under a glacier was concerned. We felt

---

76. Wild celery or cow parsnip (*Heracleum lanatum* Michx.) (Hultén 1068, p. 707); Salmonberry (*Rubus spectabilis* Pursh). (Viereck & Little 1986, p. 176–177).

somewhat skeptical, however, for we believed the Indians to be over-timorous as to water. At twenty minutes after eleven we came to the glacier where the natives said we must portage.[77] The river was very narrow, but the water did not look particularly bad as far as we could see, probably one hundred yards. Here we took a lunch, the Indians catching some salmon in a dip-net. Then Nicolai went ahead to inspect the trail over the old moraine and returned at half-past three. He said we would camp here and might be two days making the portage. This was another fearful camping-place, consisting of huge boulders, eight to ten feet through, piled over each other and overgrown with ferns and black alder. There was not enough level space for a squirrel to have stretched out comfortably. Even the Indians began grumbling, and after a consultation, they all packed up and at quarter-past-four we started over the trail to reach the river beyond the rapid, some two miles away. How we all got in at seven without two broken legs apiece is yet a mystery. Big boulders are bad enough for a heavily packed person to struggle over but when the dense fern obliterates them from sight so that each footstep is into an unknown spot unless a pole is carried to partially clear the way ahead, then one appreciates how weak the law-ful language of the land is for descriptive purpose.

Just before arriving, the Indians had a long and exciting chase after a bear, the poor brute being finally surrounded and, badly wounded, taking to the water at the foot of the swift rapids where he was probably carried against the face of the opposite glacier and dashed to death by the falling ice.

These two great glaciers—the Child's and the Miles'—both reach the river and are constantly shedding icebergs into it. The Indians are greatly afraid of the latter, not on any superstitious ground, but for two or three miles they have to navigate in front of it, and the large lakelike expansion of the river which faces it gives the waves or breakers formed by the falling ice-mountains a terrific sweep, which would ingulf any boat of modest proportions, especially their scow-like skin boats. I have seen icebergs falling from its front send geysers of water over a hundred feet into the air, the resulting tidal wave being ten to fifteen feet high, and running over a hundred yards up the beach. Fortunately the lake has a

---

77. The portage lay across the moraine upriver from Miles Glacier. At the time of Schwatka's visit this moraine nearly abutted the west side of Copper River, forming the Abercrombie Cañon and rapids noted by Lieutenant Allen.

great but slow back-current on the opposite side from the glacier and between the two currents is nearly always a dense ice-pack, which acting like a breakwater partially protects any craft under its lee. In front of the Child's glacier below, the river is narrower and swifter, and the ice seems to be of a different texture, for, instead of shedding great bergs, each explosion from the front is accompanied by a vast rolling down of powdered ice; if ice the size of boulders can be called "powder" by comparison with bergs as big as houses. The swift current and small ice both combine to prevent large tidal waves. Again, the great swiftness in its front compared with the slack water of the other, allows a crew to pass the danger in one-fourth the time, at least.

It was known from our Indians there were two or three salmon canneries in the delta of the Copper and in the Prince William Sound near by, so we were naturally very anxious to reach the comforts of civilization again, now that our exploring work was virtually over, but the state of the ice in the lake ahead of us had to be just so, before the Indians would move. A long series of explosions, lasting an hour or two, with tidal waves lashing the lake is generally followed by a like calm; but the ice from the glaciers must drift around so as to open a way back of the break; so, altogether, it is a matter of much strategic maneuvering.

All night we could hear the thundering of the two glaciers, like seacoast batteries, and the lashing of the waters on the shores. The morning of the 18th gave us a dismal prospect for getting away, and we were correspondingly depressed, for we were afraid of losing a vessel that would leave Nutchek [Nuchek] about the 20th to 22nd for Sitka, while we might even be too late for the returning salmon cannerymen for San Francisco. Before daylight that morning, the Indians had started back to bring the boat over the trail, and at quarter-past twelve, with three or four cheers, they landed it on the beach, having carried it across, upside down, on their shoulders.

The rapids around this portage were certainly formidable, and shooting them with the Atna boat would have been clearly a piece of folly, but I believe they were no worse than the same length in the Nee-zee-náh[Nizina]. At quarter-past four, Nicolai came out of the little tent the Indians had pitched on account of the rain, and, calling the Tanana chief, they held a brief consultation over the aspect of the ice. The calm had come after an hour of the most tremendous roaring from the glaciers I ever heard. It sounded as if a St. Elias avalanche had slipped down into a Rocky-Mountain thunderstorm. Nicolai gave the word to start, and in just six minutes from the order the tent was down, all the material was rolled into packs, the boat launched and loaded, and the paddles were plunged into the water as we shoot out into the lake at the foot of the rapids. Not a word was uttered except by Nicolai who, standing up on the load in the stern, gave all orders. The silence was ominous as each one bent his back and paddled with his utmost strength; all eyes intently fixed on the source of danger, the glacier to our left. About half-way over, a pistol-shot crack was heard, followed by low, suppressed "Ee! Ee! Ee's!" from our natives, but only a little berg, about the size of a small cottage, fell off, and all seemed to heave sighs of satisfaction. Once clear of the front, the Indians set up the most vociferous singing, freely interspersed with cheers, and we shot by the Child's Glacier like an arrow out of a bow, clearing its lower end at twenty minutes past five. Soon after we entered the delta of the Copper River's mouth, low, flat and marshy, cut up by interminable but swift channels and covered with willow and alder brush.

We had hoped to make the coast Indian village of Allanik [Alaganik], where Nicolai thought we might meet fishermen from the salmon canneries, but the day was so dark and lowering that darkness fell early and at a quarter-past seven we camped (No. 64) our last and worst camp of the trip, on a low, flat, boggy, quicksand island covered with water-soaked brush for firewood and into which we sank above our ankles if we remained in one spot over a minute. Our beds were made down in the mud, feebly resisted by brush-pile mattresses we gathered, and we remained awake and fought mosquitoes until we got away in the morning, the down-pouring torrents changing to seemingly solid sheets of falling water.

At half-past ten, we made Allanik [Alaganik] and routed out a lot of forlorn-looking savages of the "drowned-rat" appearance, who informed us that the Eyak cannerymen of Prince William Sound had abandoned everything and gone to San Francisco, but there were white men at George's cannery in the delta, which one Allaniker promised he would guide us to; and we got away at fifty minutes past eleven. The rain was terrible that afternoon, and when we met a little weasel or stoat swimming across a wide channel toward the boat, one member was critical enough to observe that it evidently jumped into the water to get out of the furious rain.

At four o'clock, stiff and half-frozen, we got to the cannery of the Peninsular Company, about the center of the delta (some forty miles wide) facing the coast. Captain Shaw, the superintendent, had gone but the day before, the cannery men having been sent to San Francisco with eighteen thousand cases of fish just before that. "Mike" Duvall in charge, an old Yukon prospector, who knew Russell well, with a heart as big as his body (and he weighed two hundred or over), "turned himself inside out" to make us comfortable, and completely succeeded.

There were two prospectors here, Cloudman and Boswell by name, who most generously offered to take us to Eyak in their Columbia-River boat as soon as the weather would permit, for these canneries had not yet closed, as reported at Allanik [Alaganik], or even take us to Nutchek [Nuchek] if possible. That night I listened to one of the pleasantest duets of my life—the rain falling on a roof with a soprano pitch and the old ocean surf outside the bar roaring in basso profundo.

Duvall, who had had winter charge of the cannery buildings for two or three seasons, amply confirmed Nicolai as to the winter winds, saying storms often came down the river that held him in his stoutly built house for two or three weeks at a time, and that had unroofed buildings. He told me, also, that the old Eyak chief claimed his father had seen the Miles's glacier extend across the copper River, the stream running under it. "La grippe" had swept this coast two winters before, depopulating whole villages, that were now abandoned and rotting away. Duvall told me he has seen moose on the Yukon with old hamstrung scars made by wolves, and the abundance of the former on the islands was due to the harrowing pursuits of the latter. But his most astounding statement was that he had seen a polar bear on the moraine of the Child's glacier, and insisted that he could not be mistaken as to the animal.

The 20th and 21st were hyperborean hurricanes, and it was noon of the 22nd before we got away with Cloudman and Boswell. We saw great flocks of ducks, geese, gulls, snipe and plover on the flats and adjacent waters, our course being just inside the Copper-River bar.

At six o'clock, we reached the Eyak River's mouth, where there are fish-houses belonging to the canneries of Prince William Sound, and learned the Nutchek steamer for Sitka would leave that day. We were just in time to be too late. The Eyak is a tide alternating river, so we waited until one o'clock in the morning and took the flood-tide up it. It was so high, however, with the recent rains that this lasted but an hour,

Fish house and tramway on Eyak Lake used to carry fish to Odiak on Prince William Sound near present day Cordova. (U.S. Geological Survey photograph by C. W. Hayes) [Hayes #299]

and we had to take to the oars; and at a quarter-past four, just daybreak, we passed into the lake at its head, an hour and a half seeing us across it to the upper fish-houses. From here there are two tramways, facetiously said to be the only railroad in Alaska, leading over to the two canneries on the opposite sides of the same arm or inlet from Prince William Sound.[78] Both Captains Humphrey and Mathews were absent in the Sound, superintending fishing, and I did not meet either until evening, when Captain Humphrey was the first to come in. He thought there was one chance in a dozen of catching the steamer at Nutchek, and so we took the little steam-launch *Goby* to the larger launch *Salmo*, and got away immediately for that port, at ten minutes past six. I suppose the captain would have set fire to the cannery to light us over if necessary, for when these people start in to do a favor it is never done by halves, or if it is, you get three halves and they take what is left. It

---

78. In 1891 the Pacific Steam Whaling Company and the Pacific Packing Company operated canneries at a location known as Odiak, situated near present-day Cordova (Moser 1899, p. 131).

was a rough, dark night and our second one without sleep, and I felt as soporific as if I had been sand-bagged. We reached Nutchek[79] at twenty minutes past one in the morning of the 24th, and, to our repeated whistlings, the trader came off in a *bidarka*, or skin canoe, stating the Sitka steamer had left the 22nd, but would make another trip thereto in a month. We anchored in the beautiful harbor, and I turned into the microscopic forecastle (for me) to sleep, the *Salmo* starting back hurriedly at half-past four in the morning, the captain mistaking my snoring for the ship dragging anchor. The Pacific Steam Whaling Company's cannery buildings, which Captain Humphrey superintended, were placed at our convenience until we could get away, and we were very busily engaged the next few days in taking a rest.

---

79. The village of Nuchek, now abandoned, was located on the north shore of Port Etches on Hinchinbrook Island, about thirty-five miles southwest of Cordova. In 1890 the population was reported to be 145 (Orth 1968, p. 706).

THE JOURNAL OF

CHARLES WILLARD HAYES

1891

The Journal of Charles Willard Hayes, 1891

Juneau, Monday, May 25. We are getting started, by no means a simple process. The Siwash got on a hoochinoo tear last night—nearly used up our canoe man, Robert, and this morning one of the packers is in the jug. However we have weighed our goods and find there is 1,149 pounds and we will cut it down to 1,000 before we start on the trail.

The wind is from the east, with a drizzling rain. This will take us up into the inlet in good shape. Bob didn't turn up—now he comes—and a couple of other Siwashes are missing. Finally all our plunder is loaded into a big five-ton canoe. The Indians pile in their provisions for the two weeks they will be gone, and with a rousing three cheers from the crowd assembled on the beach we are off.

There are ten in the canoe; three white men, six packers and its captain, Robert. The canoe has two big oars and all the rest of us, or eight, use paddles and we fairly skip through the water although the tide is against us as well as the wind. Near the mouth of Sheep Creek in making a spurt—racing with the steam ferry boat—a row lock came loose and we put in to repair the damage.

At 2:00 P.M. we pulled up on a kind of shingle for lunch. The Indians built a fire and made coffee and fried bacon. Started again at 2:40 P.M. At 3:35 P.M. we rounded the point into Takou inlet and hoisted two sails, wing and wing, went up flying. Driving rain all the afternoon. Numerous ice bergs are floating in the inlet, increasing in size toward the head. We had a fairly good view of Takou glacier and the glacier opposite the mouth of the Takou. The latter does not discharge but terminates in a broad flat of moraine stuff. The view was greatly obscured by clouds hiding the skyline, mountains and upper portions of the glaciers.

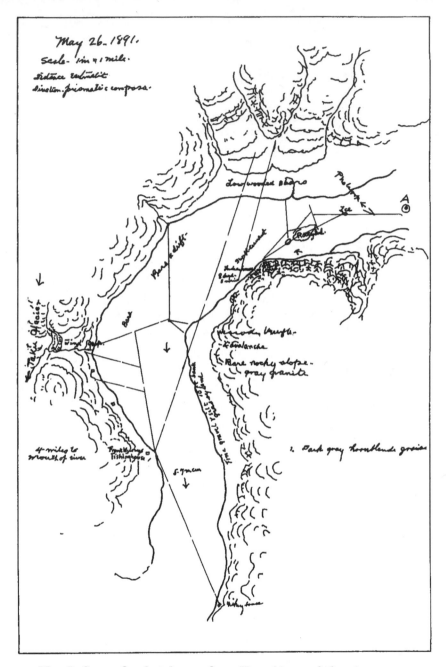

Map 2: Copy of a sketch map from Hayes' journal showing a portion of the lower Taku River in 1891. The expedition spent the night of May 25 at Frank Murry's fishing house, shown at lower left. Twin Glacier appears at upper right, the two ice-streams united at the base.

We entered the mouth of the river about seven. The river is very muddy with light grey silt, and quickly shoaled so that the paddles touched bottom and the men took to poling. We were stuck in the mud several times.

At 8:30 P.M. we reached the fishing station of Frank Murray and Company. We got permission to camp in an unfinished fish house which has a roof and a stove. The coffee is now being made. [Camp 1]

Tuesday, May 26. We spent a fine night over a loud roaring mountain stream which flows beneath the fish house. Today is a bad rainy day, but the wind is up stream.

We started from Murray's at 7:35 A.M. with a fair wind. The current, according to Murray, is five to seven miles per hour.

9:00 P.M. Camp 2: We made a good days travel today, about twenty-five miles against a strong current. The wind has been in our favor nearly all day. Toward evening we did some tall poling and paddling in the very swift current. There has been but little rain. I have not been able to get hold of a rock since leaving Murray's. There the rock is a compact massive hornblende granite. This passes into a schist at the "rapids." Above this the topography indicates schist or highly cleaved slate with cleavage vertical, weathering into extremely sharp mountain peaks, the Needle Peaks shown on our map.

We are camped on a low island covered with cottonwoods. I can't get to the mainland to see the rock, which appears to be a massive granite or possibly a very massive sandstone with vertical jointing. A cliff about one thousand feet high is on the south side of the river just opposite camp. Ahead of us appears something like horizontal stratification.

We have made a supper off hard tack, coffee and bacon and are going to bed without putting up the tent. [Camp 2]

Wednesday, May 27. We left Camp 2 at 7:30 A.M., the sky slightly cloudy. The river is braided, broken up by numerous small bars covered with willows and cottonwoods, and small sand or gravel bars covered with driftwood. The bottom is rather coarse gravel. The current is very swift and we get along slowly by hard poling. The steep rocky walls of the valley are drawing closer together. This valley has evidently been much deeper formerly and the river is flowing over a recent gravel deposit. No gravel terraces are seen on the mountain sides. Just above the two lower glaciers a well marked but shallow terrace appears on the south bluff about one hundred feet above the river. This is probably due to damming of the water by the glacier.

We took lunch on a gravel bar at 12:30 P.M. I waded across one slough, through a wooded island trying to reach the bluff to look at the rock but struck a deep swift channel and had to turn back. The bluff towers a couple of thousand feet above us, nearly perpendicular. It is evidently a bedded rock, probably sandstone or quartzite. Toward the top alternating beds of lavender and red rock occur, much contorted but with a general dip to the east.

We left lunch camp at 2:15 P.M. and spent two hours making about half a mile in very swift water, a "Skookum chuck." The boys spent a good part of the afternoon in the water, tracking, sometimes up to their arms. There are many gravel bars.

About 5:15 P.M. we saw some mountain goats on the mountain side, so decided to go into camp. We pulled into a slough of quiet water, unpacked the rifle and two Indians started after the goats.

In about twenty minutes we saw a puff of smoke and a goat came tumbling down the cliffs five hundred feet. Two half-grown kids were with the old one and the boys tried to catch them but were not able. The goat was dragged into camp badly bruised and the Indians are making a feast. We haven't tried it—too skookum. [Camp 3]

Thursday, May 28. A bright morning, not a cloud in sight. After breakfast on mountain goat and Boston beans I crossed the bayou and climbed a short distance up the mountain side. The rock is a light gray massive granite, mostly biotite and hornblende. We have evidently passed a basin of sedimentary rocks and come into another granite area.

At 11:00 A.M. we passed a rocky point on the north bank against which the main current cuts. At some distance it had the appearance of sedimentary rock but on closer examination proved to be crystalline eruptive, alternate layers of compact fine-grained granitic rock and porphyritic varieties of the same. The latter contains large (five by ten millimeters) crystals of a brown mineral, probably augite. The rocks noted on the south side of the river as breccia are also probably volcanic.

At 11:30 A.M. we stopped for lunch on a big gravel bar and took an observation for latitude. Sun's altitude 51° 28' 30".

In the afternoon the breeze sprang up and we used the sails; made much better time than by poling or paddling.

We made Camp 4 on a low island separated from high bluffs by a narrow swift channel. The island is composed of coarse gravel, with some pebbles eight inches in diameter covered with fine sand and has a sparse growth of "soap bushes" and grass among the cottonwoods. The surface is about sixteen inches above the river surface. After supper—beans and hard tack—we spread our blankets on the sand on the fresh bear tracks, and making a wickyup frame to hold the mosquito net, turned in and slipped into the sleep of the weary. [Camp 4]

Friday, May 29. Another perfectly clear day, with a slight breeze down stream. After breakfast—beans, bacon, coffee and hard tack—we waited for the wind to change. The boys are up to all sorts of tricks, racing constantly. One cut down a cottonwood tree to see the rest scamper when it fell toward camp. They have cut off a length of tree trunk and are carving a totem pole, a man's head, which they planted on the bank. We left Camp 4 at 9:45 A.M. with a moderate breeze upstream. Just above Camp 4, we passed a low rocky point, the rock eruptive, dark, compact, with some bedding and horizontal and vertical jointing.

Just below the forks of the Takou the banks close in and the river rushes through a rocky gorge about 150 yards wide. The current is very rapid and the water deep. The rocks forming the banks are limestone apparently. A short distance above the forks are bluffs of black slate or highly-cleaved shale, but we were not able to examine the rock closely.

We camped about a mile above the forks on a low sand bar. Here the river spreads out, with many bars and islands. [Camp 5]

Saturday, May 30. During the night the river rose several inches and stood this morning only a couple of inches lower than our beds. We left Camp 5 at 8:15 A.M. with a fine breeze up stream. The river is very crooked and shallow with many gravel bars so that our sails still did little good.

About one and one-half miles above Camp 5 a deep narrow valley leads west leaving a low delta of gravel at its mouth. About 12:00 we took lunch on a gravel bar opposite the mouth of valley on the east side of the river. There is a high rocky bluff with rapids at the foot, gray and blue compact limestone, much crushed, and beds contorted.

Camped—No. 6—on a low gravel bar. It took some very hard poling to reach it as the wind died out completely. The water is rising, much driftwood running. We bivouacked on the sand and built a big camp fire. Bob beached the canoe to examine damages. None were serious. [Camp 6]

Sunday, May 31. Another fine day but no wind. at 9:40 A.M. we are looking and waiting for it to come up. Bob says, "hiu skookum chuck." No hurry and will wait for wind. Without it poling is hard work and we don't get anywhere. The crew doesn't understand working the canoe in high water. Bob has yelled till his head is sick.

We left camp about noon, and with a light mountain breeze started on the last piece of the boat journey. Very swift water all the way. We had a hard fight against the current of the East Fork [Nakina River], and made the point about 7:30 P.M. Here we found two deserted, partly-ruined houses, three caches and three burial houses, the latter neatly painted, with glass windows. Inside are brass-bound painted chests (Chinese make), and a great assortment of ornaments and household goods, mostly cheap "Boston" make. Prussian glass dishes, colored wash bowls, looking glasses, china dolls, etc. We bivouacked on the point between the two forks. [Camp 7–1]

Monday, June 1. The Siwashes have announced their intention to rest a day before starting on the trail, so we are in no hurry about getting up. We lie until the sun gets too hot, about 7:30 A.M., and decide to take a tramp.

At 9:30 A.M. we start for the mountain to the northwest. Our esti-mates of its height vary from 1,700 to 1,900 feet. It proves to be 3,070. We found the brush very bad even where it had looked quite open from below. From two thousand feet upward we found much snow in the woods, about three feet thick, thawed to the ground around the trees, making sort of craters; very hard walking. It began raining about five hundred feet from the top, a very steep slope of sand and gravel, evidently glacial material—granite and basalt boulders to top 3,070 feet above the valley—later corrected to 3,100 feet.

We got back to camp about 6:30 P.M. and found the packers kicking. According to our original agreement two were to have only one pack while the others had two each, as we thought all the stuff would go into ten packs. The two one-pack Siwashes objected to the arrangement, wanting all the same. The matter was compromised by making twelve packs on condition that the packers secure an Indian who had been loafing around camp all day to go along as a guide. They had a long squabble with him trying to get him to go for $12, he asking $15, but the war was transferred into the enemies country so we were not disturbed. The packs were finally apportioned in such a manner as to give moderate satisfaction all around. It cleared up so completely that we went to bed with no protection but a mosquito net as usual. There was a hard white frost during the night. [Camp 7–2]

Tuesday, June 2. The packers started off with their first loads at 5:30 A.M. We got up later and were eating breakfast when they all came filing back into camp. It was evident something was wrong. They had gone two miles to the camp of the Siwash who has agreed to guide us till we could see the lake for $12. and he had gone back on his promise. After a great deal of talking, Robert offered to go over as guide for $10. Not for the money, as he assured us, but because he thought a great deal of us and wanted to see us safely through. He is very anxious to have one of the pictures of himself and canoe which we have taken, and to have a "Hi-u skookum paper."

I was focusing the large camera a couple of days ago when the Indians gathered around and wanted to look in. I allowed each in turn to put his head under the cloth, causing unlimited astonishment.

Finally we got the second loads allotted and at 9:00 A.M. we left Camp 7. Our course is up the side stream toward the north east. The valley is narrow and the stream flows in a single channel, very rapid. The sides of the valley are rather gently sloping and fairly well but not densely wooded—cottonwoods or aspen and fir or spruce. The valley as seen from above strongly resembles one of the valleys in the mountains of northeast Georgia, except the dark green of the rhododendron is wanting and the timber is smaller. Only a few peaks can be seen which go above timber line, here about thirty-five hundred feet above the river valley.

The packers have made very poor time today. It has been hot, and in places the trail is blocked by fallen timber. Half of our stuff is at the "Nubbins," four and one-half miles from the start, and we are waiting while they go back two miles and bring up the rest. We will probably camp here tonight. We are on a saddle two hundred feet above the river between the west "nubbin" and the mountain, commanding a fine view up and down the valley. [Camp 8]

Wednesday, June 3. The boys got away with the first load of packs about five this morning. I started about 8:30 A.M. leaving Schwatka and Russell behind with the second load to wait the packer's return.

Above Camp 8 the valley narrows considerably and the stream flows along the base of cliffs on one side or the other. It does not flow on bed rock yet but on gravel deposited in a former deeper channel. About two-thirds mile above Camp 8 is the first exposure of granite for a long distance; light gray, slightly gneissoid black mica and hornblende. The trail is very rough, up and down over the rocky points that jut out to the river.

We reached an Indian village, not inhabited, at 11:50 A.M. and found the packers asleep. I opened the instrument box in a hurry and got a fairly good observation on the sun. It would have been better but for the wind disturbing the mercury of the horizon. The gray granite extends to about 7 mi.+400[1] where talcose slate is seen extending to the Indian houses at 8 mi.+100.

At 1:30 P.M. I made two exposures with the large camera looking up river with a group of Indians and the Indian house in the foreground. The first was probably not good as the diaphragm was partly out. Second value: 8.

About 8:00 P.M. I went down to the point on which the graveyard stands and took a view up river; lost a screw for attaching the camera to the tripod.

I bought a pair of moccasins ($.50) fresh salmon ($.50) and squirrel skin robe ($3.00) off the Indians. A little girl about five years old was packing about a twenty-pound pack by a strap over her forehead and one over her breast.

---

1. For convenience Hayes recorded distances in miles and tens of feet. The notation 7 mi.+400 indicates a distance of seven miles plus 4,000 feet, or roughly seven and four-fifths miles from his point of beginning. In making his track survey Hayes estimated distances by eye and by pacing, with forty paces equalling about 100 feet on level ground. According to Hayes an experienced pacer should be able to estimate distances within two percent on level ground and ten percent on the steepest slopes. (Hayes and Paige 1921, p. 19).

We broiled a piece of salmon for lunch, no salt, rather fresh. The rear column came up a little after 8:00 A.M. and we had a good supper on salmon and a grouse which Robert shot in the morning.

It rained a few drops in the afternoon but cleared up completely about nine so we bivouacked as usual. [Camp 9]

Thursday, June 4. The packers started this morning about 6:00 A.M. We had a rather late breakfast and got off about 9:00 A.M. Met the boys coming back about a mile out of camp. We found a good trail with no fallen timber on the steep climb to an Indian village [Klik—noo].

At 10 mi.+275 are two houses of upright posts, five caches and one grave house. The houses have recently been inhabited but no one is here now. The houses are just below heavy rapids. The river makes a sharp turn and breaks through a barrier formed of big rocks, twenty to forty-feet in diameter, which have tumbled down from a cliff on the west side. The fall at the rapids, within 150 yards, is probably eight feet.

I saw Robert on the rocks with a salmon hook trying to catch a fish in the eddies below the falls. This is where the salmon we had yesterday was caught.

It is 11:45 A.M. and the boys have come in with the second load of packs, all except "The Kid" who is having rather a hard time of it. Here he comes now smiling as usual. After throwing off their packs they rush down to the river and wade in up to their knees with moccasins on and stand in the ice-cold water several minutes.

After lunch we got on the wrong trail and went 600 feet up the side of the cañon. Had a good view.

We camped at twelve miles, where the river forks. A roaring torrent [Silver Salmon River] comes in from the north around the hill at the forks. [Camp 10–1]

The sky clouded up toward night and a drizzle began about nine. We were too smart to put up the tent, and towards morning water ran under the edge and got us wet. The steady drizzle lasted all night. We spread the tent over us in the morning and later put it up. We got breakfast about 10:00 A.M., the boys grumpy. Toward noon the clouds lifted a little and the boys started with the packs. The afternoon brought clear sky and things got fairly dry. Russell and I took our guns and hiked up the mountain side, a very steep tower of limestone, to look at a red stain. The stain is produced by red clay cementing the fragments of a limestone breccia. This may be connected in some way with the intrusion of eruptive rock in the valley west of the big white mountain. The limestone is light bluish-gray, quite pure, just such as would produce red soil by weathering in the South.

We found no game and got back about 6:00 P.M. The creek had raised during our absence. Logs were jammed in above the bridge and crossing was fearsome in the middle of the stream. Our bridge washed away after Russell crossed, so we fixed up a new one. I took a picture of the camp, etc. The bacon bag went with the first pack so we can't boil any beans. We made a supper and breakfast on coffee and crackers. [Camp 10–2]

Saturday, June 6. We got a fairly early start shortly after 6:00 A.M. this morning, and found that part of the outfit was cached nearly five miles from camp. We continued on the same trail at first but soon decided it was the wrong one. Schwatka and Russell back-tracked while I climbed up the side of a very steep white mountain. About 1,300 feet above the river I found the trail we had missed, and two Indians. Had lunch there. Don't know where the rest are. The steep mountain slope is limestone with irregularities filled and smoothed over with sand and gravel. From this elevation the country to the northwest appears to be a rolling plateau 2,000 to 3,000 feet above sea level with rounded, gently sloping mountains rising a few thousand feet higher. The rivers flow in deep narrow valleys with western slopes except in a few places where the slope is to the south.

We camped, June 6, on the mountain side along the trail, the rushing stream 1,300 feet below. The slope was steep and we had to build up and dig out places for our beds. [Camp 11]

Sunday, June 7. Today we made a long march, the first two miles over a bad trail. It was wet and swampy in the woods, with much snow on the north side of the big white mountain. At about 19 mi.+250 we came out into an open valley and found a stream which had been hidden in the cañon below. The vegetation was thick moose brush, which looked exactly like cultivated current bushes, the leaves just beginning to come out. I lost the trail, thought it crossed the creek, tried to wade and got into a hole with my boots full of water. Coming back I got them full again, built a fire and took my clothes off to dry. Russell saw the smoke and came over from the trail. We joined Schwatka at twenty miles. From this flat to Camp 12 the trail leads through a broad meadow with few trees, the trees growing thicker toward the foot of the mountains on each side and running up the mountain sides three hundred to five hundred feet above the valley. The surface is covered with boulders, mostly granite with large hornblende crystals and is evidently an old valley filled with glacial debris.

We found the first packs cached at 22 mi.+250. A couple of Indians came up shortly, faces blacked, dogs packing the kitchen outfit. After Schwatka came up we got out our fishing tackle and went along a low ridge to the lower end of the lake [Katina Lake], a mile long, one-fourth mile east of camp, and tried fishing with flies but no luck. We met a Siwash with a big string of trout which he had caught with a hook on the end of a pole in the stream below. We got some salmon eggs for bait and in about half an hour caught nine trout. Four were beauties, twelve to fourteen inches. We had a hard scramble to get the fish after landing them, and got wet again. The second packs got in about 7:30 P.M. and we had a feast on fresh fish. [Camp 12]

Monday, June 8. Our packers kick; they are out of grub and won't go on. They blame Robert for telling them they would be back in Juneau in two weeks from the start. They decide to send Paddy back to the canoe for more grub. Schwatka agrees to feed the rest till his return. They divide Paddy's packs among the rest. Robert decides to go back in spite of his promise to guide the boys through. About one and one-half miles out we met a large company of Indians from the Interior bound for Juneau. The group included three men, seven women (one old woman in a pack), three babies, and five pack dogs. Dogs and children were packing. One little boy of about six years was packing skins weighing about forty pounds. I took his picture. Their faces are nearly all blackened, even the babies. The chief had three wives, one old and two young. One old woman was in a pack carried by a young one. She resembled a mummy, very old, with a labret in her lower lip. The squaws came up later and camped near us for the night.

The boys got a fresh supply of moccasins and buckskin for soles. They spent the morning about the camp fire mending foot gear. They made short trips today with each pack, going back for the second and resting on the way.

The Indians advised the boys to go around the mountain which lies between the two forks of this upper branch of the Takou in order to avoid the snow. After crossing the first west fork we left the main summer trail which leads over the mountain about N45°E and followed the tracks of the Indians. It is difficult to keep the track. There is no well-marked trail and moccasins make very little impression on damp moss. Up to Camp 13 we traveled across meadows of water-soaked moss, frozen at a depth of from six to eight inches. Wherever there is a slight rise, a gravel and boulder ridge, there is a dense growth of spruce draped with black moss.

Camped, No. 13, on a slight knoll. I took the rifle and went out to see if by some accident I could see some game. The moose tracks are quite abundant in the mossy meadows and apparently fresh but so far no animal has been seen. Game of any sort is extremely scarce. We have had a grouse nearly every day on the trail, never two the same day. The common name here is "fool hen," a little smaller than ruffed grouse and light gray. When flushed it usually flies up on the nearest tree and looks at you. Mosquitoes are bad. [Camp 13]

Tuesday, June 9. We had a slight shower in the night and several in the course of the day. Before starting several signal smokes were sent up for Paddy. The method of signaling is to set fire to an isolated moss-covered spruce tree. In a few seconds the whole tree is wrapped in flame with dense clouds of black smoke.

We fired trees at intervals during the day. Most of the forenoon the trail was over moss meadows; hard walking. After noon we struck the edge of burnt timber; more than one hundred square miles have been killed. Fairly good walking.

While waiting for the packers I took the shotgun and went up toward the high mountain, north to a round knob about 700 feet above the trail. I had a good view of the country to the east, across the range separating Takou drainage from Hootalinqua, and saw a high range of sharpened mountains, and some very high peaks to the west.

The knob and range northwest of the trail is composed of gray granite containing black mica, clear transparent feldspar weathering opaque, white quartz, and large crystals of hornblende. This is evidently the source of the boulders which I saw so abundantly in the valley of "Meadow Pass." The surface of the granite is rounded into moutonnee, but no scratches were seen.

Coming back I found camp on a small ridge near a boggy moss meadow. Paddy had gotten in a few minutes before me and was given a rousing welcome. The boys have been dumpish all day but are "hias close tum tum" [in Chinook jargon, "very happy"] now and are laughing and joking again. This is a good sign. The mosquitos are very thick but a frost at night makes them quiet so we are not disturbed in the morning. [Camp 14]

Wednesday, June 10. The boys got an early start this morning about 3:00 A.M. and were back for their second load before we rose. The trail is very dim through the burnt timber. I stayed behind to pick up my line of last night and have some tall trailing to find the way.

We made about six miles today and camped on a low ridge in the burnt district. The ground is very swampy. All arrive and the mosquitoes are very bad. In the valley there are many long oval ridges ten to fifty feet high, sloping off at both sides, their long axes in the direction of the valley. Some appear to have rock in places but most are made up of gravel, sand and boulders. John treed a hen just before we stopped and I shot it. I gave him a piece of bacon for it. [Camp 15]

Thursday, June 11. We continued all forenoon through burnt timber, the boys doing their best to make the river. We came in sight of it about noon. It flows in from an open valley, the channel slightly depressed with swampy meadows along the sides, a bed of gravel and boulders forty to fifty feet wide, three to five feet deep, current five miles per hour. The water is clear, slightly brownish. We tried fishing but there was no sign of fish in the stream.

The trail heads up stream about two miles to the crossing. We went ahead to await the packs. It is a hot day, the mosquitos very bad. The packers came in about 7:00 P.M. having made seven point two miles today, the best day's work yet. The boys are out of grub again. Paddy brought only about twenty pounds when he went back. [Camp 16]

Friday, June 12. We set up a boat last night as the raft on which the party of Indians we met crossed was not to be found and there is no timber to build one on our side of the creek. It took just half an hour to set up the boat.

This morning Russell ferried the packers across with all packs and our bedding. The trail heads up a side stream valley about one and one-half miles wide, sloping gently down the stream. There is a thick forest of spruce with moose brush about two miles up from the main valley; very thick and hard traveling, with a poor trail. The moose brush is evidently an alder. It grows six to eight feet high in the lower valleys and decreases in size to six or eight inches on the mountains above timber line.

The trail became very dim as there was snow on the ground when the Indians passed. Finally we lost it altogether. The boys gave up and we camped alongside the stream. After supper Russell and Barney went out three miles ahead and looked over the country. They killed three ptarmigan on the way, and Russell had killed three grouse during the day. The mosquitos are very bad. We have to build a fire the moment we stop. They crawl through our netting at night. [Camp 17]

Saturday, June 13. The boys were very dumpy this morning. Russell went along and talked a little "tumtum" [spirit] into them. Schwatka and I got breakfast and followed later. We crossed the creek on a bridge Russell made in the morning and got into better walking. The timber is scattered and the moose brush not so thick. We climbed up on the south side of the valley by a gentle ascent to timber line, and saw a curious form of dwarf pine, mosquitos abominable. Saw many ptarmigan. We reached a knob and beyond it a cache of the first packs about noon. Rain beginning, very heavy thunder and lightning and the ground soon covered with hail. After it cleared off I took a shotgun and started west to a line of knobs. Shot one ptarmigan near camp. The knobs are composed of gray hornblende granite, very similar to that seen on the west of the valley. The rock immediately in the valley we traversed this morning is a dark silicious slate. We reached the top of the knob at 4:00 P.M. and have a fine view to the East, with the Big Lake [Teslin] occupying the valley to the north while the continuation of its valley to the south is heavily timbered and filled with small lakes. I counted fifty-four lakes plus the big one, and the upper end of that west country some distance south of it is concluded by another knob. High sharp mountains lie to the east of this great valley. I shot three more ptarmigan. The second packs came in about seven. A cold wind rose about 9:00 P.M. and quieted the mosquitoes, which are vile here. Camp is just at the foot of a snow bank reaching to the top of the mountain, a cold stream flowing from the foot. John went up the side of the mountain to shoot a woodchuck but it turned out to be only a rock. I saw a couple in the afternoon. I graduated a protraction after 9:00 P.M.; it is very light. [Camp 18]

Sunday, June 14. The boys got off early between 2:00 and 3:00 A.M. with the first packs. We started about 6:00 and continued three miles across a rolling basin to a gap between low hills. There is no timber, only low willows and moose brush, moss and some bunch grass on the knolls; hummocks, round patches surrounded by moss one to five feet across, quagmires, with loose stones on top.

At the pass, 58 mi.+250, I found white vitreous quartzite in place, black slate ribbons in place near. There was ice on the two lakes of this upper basin. We built a fire in the moss to clear off mosquitoes. It sparked badly and burned about an inch off from the top. We got down to flat land near the lakes and camped in a swamp, the mosquitoes bad. We tried fishing but the lake was too high. [Camp 19]

Monday, June 15. We made an early start soon after sunup at 3:00 A.M. Packed a mile to the first lake, put up one boat and loaded all our outfit into it. Schwatka and I paddled down two and one-half miles, to the lower end, while Russell and the Indians walked. We made another portage to a small lake, repeated the same process, and camped at the lower end of the fourth lake near its outlet. There was much vegetable matter in the water. The lake has a small outlet and probably does not discharge in the summer as it gets no water from the mountain. The mosquitoes were worse than ever. Schwatka and I put our beds together and doubled the nets but the 'skeets came through all the same. [Camp 20]

Tuesday, June 16. We got off about 6:00 A.M. and made a portage of about one mile to a small lake. At the lower end we landed on a beaver dam. Another short portage, and coming up on a low ridge we saw Lake Aklene [Teslin] spread out before us. We reached its shore at 10:00 A.M. and put up the tent. A light rain and good breeze kept the 'skeets off. After dinner the boys were paid off, with fairly good satisfaction. About three Russell and I took them a couple of miles in the boat up a "bight" and put them on the trail. It was a great relief to see them disappear. We went over to an old saw pit and got slabs to make oars, and were run out by the 'skeets. [Camp 21–1]

Wednesday, June 17. We spent the forenoon on the oars and boats. Schwatka got an observation for latitude. After lunch Schwatka and I rowed the *Frederick Bonner* up about four miles toward the end of the lake and mapped as far as we could see. Coming back we had dinner and then put up the second boat, rigged masts, and sails of slickers. Rain began about 9:00 P.M. and it rained some all night. [Camp 21–2]

Thursday, June 18. We got up at 6:00 A.M., finished rigging the boats, packed up, and at 9:00 set sail. We are now scudding down the lake before a good breeze making over three miles an hour. Russell is in the *Frederick Bonner* and Schwatka and I in the *Robert Bonner*. We are well rid of 'skeets for once, and it is a blessed relief. My face, neck and hands are swelled up, lumpy and raw. The sky is cloudy and the wind quite cool.

11:15 A.M. The eastern shore of the lake is low and heavily wooded—spruce and cottonwood—with only a few low ridges, probably gravel, less than one hundred feet high back two or three, probably five miles, to the base of the mountains. All the mountains that can be seen from the lake are rounded somewhat, timbered entirely to the top and probably not more than two thousand feet above the lake, closely resembling some parts of the Appalachians, e.g., northwest Georgia. West of the lake the country is low for two or three miles back with low ridges to the foot of rounded, wooded ridges 1,500 to 2,000 feet high. Back of these five to ten miles is a range of sharp peaks with some snow.

Rocks exposed on the west side of the lake at 16 miles (from Camp 21) are dark greenish-black, fine grained, compact. They break in rhomboidal fragments with sharp splintery edges, probably a basic eruptive.

At sixteen miles from Camp 21 a high rocky headland juts out into the lake. The rock is a coarse-grained granite, with feldspar in large crystals, almost porphyritic, light pink. Mica is greenish, much changed as well as feldspar. No hornblende was seen.

We camped at about twenty-three miles on the point of a peninsula on the east side of the lake. The lake is very high; Russell says six or seven feet above its level of a year ago. Water comes up over all the beaches and into the timber. Fringes of drift along the shore makes landing difficult. The point is composed of sand and gravel with many larger boulders. The same holds for all points seen along the lake from Camp 21. Boulders and sand seem mixed indiscriminately. The 'skeets are not so bad as usual. [Camp 22]

Friday, June 19. We slept well without putting up the tent. 'Skeets didn't get through the net. We got up about 6:00 A.M. It is a cloudy morning, a light wind down the lake. We started at 7:00 and the wind died out immediately. Rusell went to a small rocky island for gull's eggs; he got two. At noon we put the two boats together for lunch but didn't land on account of mosquitoes. The surface of the water is glossy. Lots of fine salmon trout come to the surface but we can't get a bite.

In the afternoon a rain storm came up from the southwest, but passed east of the lake. We had a fair wind for half an hour and then it turned suddenly and blew hard. We were in the middle of the lake and had a hard pull to the east shore for shelter. We tied to a submerged cottonwood which broke the wind somewhat and rode it out. The waves quieted down quickly and we pulled out. I hooked a salmon of about eight pounds and landed him after ten minutes. Later Russell caught one weighing about twelve pounds. We had one for supper.

We crossed a deep inlet [Morley Bay] on the east side of the lake, into which a river—ascended by Russell last spring—empties. It is larger than either branch of the Taku and flows from the south parallel with the lake and behind the first range. The valley as seen from the inlet is broad with gentle slopes, and wooded. The eastern shore of the lake is a bluff fifty to one hundred feet high, of boulders, clay, gravel and sand, the lake cutting at some points. No stratification was seen. There seems to be a very gentle slope from the edge of the bluff back several miles to the foot of the ridges. The bluff occurs at places on the west side of the lake, though the slope generally is even from the water's edge.

We camped at the foot of a bluff on a little level land covered with small cottonwoods and willows, only a few inches above the surface of the lake. The beach was covered. Here we saw the first mosquitoes since leaving Meadow Pass. [Camp 23]

Saturday, June 20. There were no 'skeets last night to speak of and life is again worth living. We breakfasted on fresh salmon, gull's eggs, beans, bread and coffee, and started at 6:55 A.M. with a fair breeze, sailing about three miles per hour.

At 7:00 A.M. I got out on a sandy point and took a photograph of the Caribou Mountains. Getting into the water I broke the compass glass.

At 8:00 A.M. we reached the mouth of a big river, or bay, into which it empties [Nisutlin Bay]. There is a sharp line between the dark blue water of the lake and the muddy water of the river. At 9:30 A.M. we began rowing and at 10:00 heard a gun on the west side of the lake and saw a couple of Siwashes coming out in a canoe.

The Indians were Takous in a spruce bark canoe, the bark inside-out, sewed to a rude frame and caulked with spruce gum. They had moose meat and wanted four dollars for a piece but came down to one, and then paid that to Russell for part of a bag of smoking tobacco. They could speak a few words of Chinook and English. We made a low sandy point and I got a series of sextant readings for latitude. We lunched here on fried moose meat; very few 'skeets.

After lunch I took a couple of pictures with the Siwashes in the foreground. We sailed away with a light breeze which soon died out and rowed the rest of the afternoon. I hooked a fine salmon trout but the hook broke and I lost him.

We camped about 6:00 P.M. on a little flat ground at the foot of a bluff. Had a time getting through the drift wood to land. Our camp is in a dense growth of spruce trees of peculiar tapering form, and a deep carpet of moss covers the ground. Robins sing when the wind is still. It is still light after 10:00 P.M. The mosquitoes are worse than at the last camp, probably on account of the moss. [Camp 24]

Sunday, June 21. We slept late this morning waiting for a fair wind. Breakfasted on boiled moose meat, bread and coffee, and got away at 8:50 A.M. with a light head wind.

We rowed all day against the wind, not landing for lunch, and reached the lower end of the lake about 5:00 P.M. We pulled through brush at three-fourths mile below the lake for camping, a poor place, with the mosquitoes rather bad. Decided to stay until noon for an observation if the weather proves to be good. [Camp 25]

Monday, June 22. We slept 'till 8:00 A.M. The sky was cloudy to the south and so decided to move on. We started at 9:10 A.M. with a rapid current for a mile (about four miles per hour) then slower (three miles).

The river swings back and forth between bluffs three-fourths to one-mile apart, composed of very fine stratified sand. The first bluff is about one hundred feet high from the flood plain. Back from its edge one to five rods is a second terrace thirty feet high. Where the river cuts the bluff sand comes down in avalanches sending up clouds of dust. The sand is probably frozen below the surface. In some places the river spreads out from bluff to bluff, lake-like sections with slack current. The valley has about the same width and general character as when occupied by the lake; eight to twelve miles broad, to the base of rounded spurs from a high plateau on the western side.

On examination the bluffs proved to be composed of fine silty sand, much cross-braided with thin (one to three-inch) seams of blue clay. Four miles further down, at nine miles, on the east bank are layers of sand some fourteen to sixteen-inches thick separated by thin (two-inch) layers of clay. The sand contains more clay, the strata contorted by slides. Near the base of the bluffs for ten feet above the river the surface is covered with a white crystalline efflorescence, probably alkaline carbonate.

It has been hot all morning. At 1:00 P.M. a very heavy thunder storm came up from south. The shower passed east of us, hanging in the valley to the northwest nearly all day, and a cold wind from the north blew all afternoon. We struck the narrows and swift water toward evening. The river is about 120 yards wide with current about ten miles per hour. We swung in behind a point in an eddy and camped on deep moss; a very wet evening. [Camp 26]

Tuesday, June 23. We started about 7:00 A.M., the water swift and the river very erratic. About noon we had the boats tied together when we struck rapids and waves dashed over the boats, getting both wet, including the note book. We stopped on a low sand bar to dry out and eat lunch just below a good-sized creek coming in on the west side, a cold wind blowing from the north. In the afternoon we saw the first sign of a volcanic-ash stratum, fairly continuous from here on. The sand bluffs are capped by gravel, the gravel gradually replacing sand.

We camped—No. 27—on a low shore a couple of feet above water among a few small, dead cottonwoods. The site had been burnt over, and made a very good camp. After supper I took the shotgun and went up on the terraces, the first about thirty feet and second one hundred feet above the river. The original level was of fine sand and gravel with a few cottonwoods, mostly burnt-over. I saw a lynx in the bushes but failed to get a shot.

Yesterday I shot twice at ducks and twice at geese and didn't get anything. Today Russell shot twice at geese and once at ducks and got the same. Both disgusted. [Camp 27]

Wednesday, June 24. We slept late this morning; very few mosquitoes. Got away at 8:37 A.M. in swift current. We landed at 1:00 P.M. on an island at an old bar—workings all around by high water.

At 2:40 P.M. Russell saw a moose on the east side of the river. He snapped at it but there were no cartridges in the gun. He threw one in and got a shot. The moose stopped and he shot again. By this time the current had carried us downstream. We paddled across to the lower end of the island on which the moose was seen. Russell got out and said there was no moose, but that he had seen a couple disappear up the island and swim across to the mainland. He followed them up but failed to get a shot. Coming back he found the one first shot at dead. We pulled the boats up to the spot and found a fine young two-year-old buck. I took a picture of Russell and the moose. We took the hind quarters, liver, heart and nose. It would dress at probably three hundred to four hundred pounds. We got away at 4:00 P.M.

We found the junction of the Hootalinqua [Teslin] and Yukon Rivers about a mile below. The water of Hootalinqua has been getting more and more charged with sediment since leaving the lake, and we can hear the metallic clink of gravel rolled along the bottom. The Yukon water is clear blue, and does not appear to be nearly the volume of the Hootalinqua.

We ran along with a good current at about six miles per-hour until 5:45 P.M., when we saw smoke on the west bank. We pulled in and landed a short distance below the camp of two miners, Swingler and Harvie,[2] and their Indian helper. They are waiting for the water to go down so they can wash gravel. We decided to camp here. Had a feast of fresh moose meat, and roasted the nose for a later lunch. It was fine eating. [Camp 28]

Thursday, June 25. After breakfast I took a photograph of the miners and their camp, including their much-troubled dog, "Boodle," and their boat the "Creeping Jesus."

We got away about eight. At Lower Cassiar Bar we found a solitary miner, Monahan, cooking his breakfast. He had cleaned off the sand and had a pile of logs burning to thaw out the gravel. We stopped a few minutes and then went on. A few miles below we saw a camp on the west side of the river but no one at home, and didn't stop. We crossed the river and below the next bend found another camp with a couple of fellows, Jones and Harrington, and Cummings and Dawson from the camp above, waiting for the water to go down so they can work the bar. Cummings was Russell's partner last year. We stopped for half an hour, and gave a quarter of moose to the four miners. About eight miles below on the west side was another camp with three young miners, King, Brinton and Trimbly,[3] also waiting for the water to fall. We ate lunch here.

At 6:25 P.M. we passed the mouth of Little Salmon River and camped about a mile below. Little Salmon is high and nearly as muddy as the Yukon, and extremely crooked. After supper, about 8:00 P.M., I went up a small mountain west of camp. The slopes are smooth, with a little grass and a kind of sage plant.[4] Higher up on gentle slopes were small

---

2. Schwatka recorded the names "Swingler" as "Swindler" and "Harvie" as "Harvey."

3. Schwatka spelled the names "Cummings" as "Cummins" and "Trimbly" as "Tremblé."

4. One of two species of sagebrush found in interior Alaska and Yukon; *Artemisia frigida* (Willd.) or *Artemisia alaskana* Rydb. (Viereck & Little 1986, pp. 252–253).

cottonwoods and some spruce. The rock is a very coarse conglomerate, highly indurated, with jointing planes directly across pebbles. Bedding is indeterminate, the strike is with the ridge, and the dip probably N65°-70°W. From the top, at 1,350 feet above the river, I had a fine view of very distant white peaks to the southeast, a few other patches of snow—no higher mountain—rolling ridges with smooth outlines. A velvet green covering over everything, open grassy parks, cottonwoods and spruce in contrasting greens. The narrow valley of the Little Salmon is very crooked in its lower course. The wide valley, N20°W, is apparently a continuation of the Yukon valley. "Rosebud Butte" separates it from Yukon River. It is occupied by a number of lakes and drained by a small stream flowing into the Little Salmon a mile or so above its mouth.

There was a large variety of flowers on top of the mountain and clouds of 'skeets. A strong wind was blowing and when it went down I couldn't stand them. There were very few 'skeets at camp on the river bank although a swamp was just rods back of it. I got back to camp at 10:00 P.M.—still light. [Camp 29]

Friday, June 26. We left camp at 6:40 A.M. The current is stronger than yesterday, probably between five and six miles per hour. There were few marked features along the river except Eagles Nest, a high rounded hill of limestone which we floated past at 12:15 P.M. while eating lunch. At 1:00 P.M. we passed the mouth of Nordenskiold River, a very insignificant looking stream at the mouth, but with a rather wide valley. Marks of ice action are abundant along the river, the banks ploughed up and heaps of gravel piled up.

About 4:00 P.M. we came in sight of Rink Rapids [Five-Finger Rapids] and we pulled into an eddy a short distance above. Russell and I went down on the east bank to inspect. The effect of high water is to make the rapids rougher than usual. We decided to run the east channel. Russell started ahead with oars, facing the bow. Schwatka and I followed with paddles. Russell got out of the swiftest current and we overtook him. After passing the second wave his oar slipped out and his boat turned side-on in the trough. Our bow hit his boat and we expected that the next time we came down it would be right on top of him, but he succeeded in getting his oar in and turning completely around. He shipped some water, a pint or so, and but for the excellence of the boat would have swamped. We didn't take a drop of water. The waves were four to six feet high and some curling.

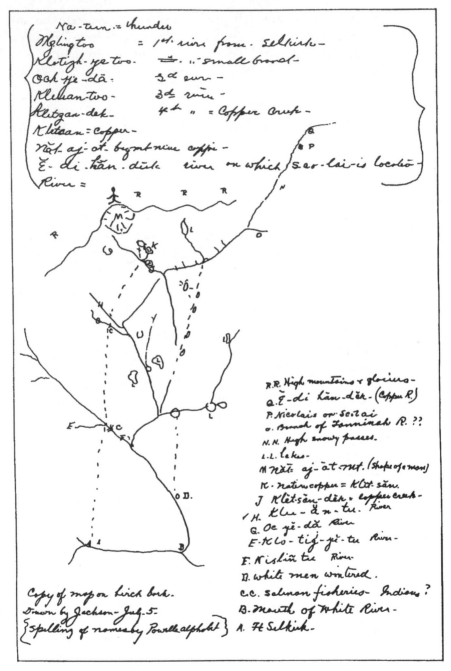

Map 3: Copy of a sketch map from Hayes' journal showing the source of copper long known to the Natives of Alaska and Yukon. The original map, drawn on birch-bark by Johnson, was copied by Hayes.

About an hour later we saw heavy water ahead [Rink Rapids]. We landed on the east side and Russell and I went down to see. We found smooth water on the east side which we ran without difficulty.

We camped on a grassy bench six feet above the water; saw several rabbits, the first I have seen. Moose meat saved them. [Camp 30]

Saturday, June 27. We got away about 6:30 A.M., with showers all forenoon. Shortly after noon we saw the mouth of Pelly River. We were looking for Harper's store on the point between the two rivers but soon saw it on the site of the old Selkirk House, near its old chimneys which have been pulled down. We crossed over, and found that Mr. Harper had gone down the river to St. Michaels leaving things in charge of Sam, an Indian from "Charley's Village" down the river, who can talk some English, and two young Stick Indians who are working for their board. We ate our lunch and then had a powwow with the Indians. Sam knows very little of the country about here and interprets only indifferently as he knows but little of the Stick language. One of the Stick boys appears to be a bright fellow. He drew a map of White River that shows the location of copper, where they make copper bullets. He says it will take eight days with packs to reach it. We got the boys to go a short distance up Pelly River to tell the Indians camped there to come down. They started about 4:00 P.M. We promised them each a butcher knife if they bring some Indians back.

Harper's house, near which we are camped, is on a terrace of coarse gravel and sand about ten feet above the present surface of the river. It is a good-sized log house, comfortably fixed. There are several garden patches planted in potatoes, beets and onions. The potatoes are up three to five inches and look well. All the trees have been cut for some distance back from the river and the mosquitoes are quite uncommon. We spent the evening washing and patching. [Camp 31–1]

Sunday, June 28. Light showers all morning. After dinner Russell and I rowed across the river to see the missionary's house. There is a strong current. The water on the east side from Pelly is muddier than on the west. It was hard work getting across above a bad drift on the head of an island. We landed on the east side of the river at a low bank four feet above water, with heavy vegetation and the largest spruce trees I have seen, some sixteen inches through. The mosquitoes were very bad. There are several small cabins. No wonder the missionaries went crazy in such a place. We tried to get back to the basalt cliff but after crossing a couple of slews came to a deep channel of running water and had to turn back.

On recrossing the river we had the benefit of the current and landed easily a few rods below camp. The boys had come back on a raft. They reported a camp up the Pelly, mostly squaws.

We did another job of washing this afternoon. About 4:00 P.M. a raft came out of the Pelly, the Indians pulling for dear life for the western bank. They landed a short distance above us and went into camp. The party consists of a middle-aged buck, Don-ghe = Johnson, sick with rheumatism in his legs—very sick till he found he couldn't get anything out of us by begging—and Mrs. Don-ghe, a hideous old squaw with a big silver ring in her nose and her chin tattooed with several vertical lines from her mouth down. All the old squaws I have seen are marked in this way. None have faces blacked like the Takus. Mrs. Don-ghe was horribly sick with a pain in the stomach. I gave her a strong dose of Jamaica Ginger, and she hasn't been about camp since.

Next was a skookum young squaw (Matilda Featherlegs) who had no complaint but bashfulness and who did the work for the crowd. Next were three boys (see picture) the oldest about fifteen, Iz-Yum, an orphan. Next Mug-Ich-Luk = Harper, about nine, and last Peter. He has no Indian name as the missionary caught him very young and baptized him. He is about five. We call him Mulligatony [sic], as he is generally in the soup. The last number of the family is a little girl about three.

In the evening Jackson came down and told us all about the country between here and Copper River. Even the kids seem familiar with it, which is encouraging. Sam brought out four bullets and a slug made of copper; Schwatka bought them. Jackson promised tomorrow to send up the river for Indians whose smoke he had seen. [Camp 31–2]

Monday, June 29. Don-ghe didn't send after the Indians as he promised. With a little coaxing he went himself at noon but came back in a few hours without having seen anything but a moose track. About nine in the evening the two boys started out. [Camp 31–3]

Tuesday, June 30. Sunday I got a dressed deer skin from Sam and yesterday and this forenoon have made a pair of pants. I cut them out by my black ones which are about gone. They are a fine fit and the admiration of all, even the Indians. The latter are trying to tell me something about smoking them which I don't quite understand.

Shortly after noon two rafts appeared coming out of the Pelly. They were going down on the other side but we signaled to them and they pulled across, landing about one-half mile below us.

Soon two more rafts appeared from the Pelly and they also landed. All have gone into camp down there, about twenty old and young, and a miserable filthy set they are. We will probably be able to get three or four packers from the lot. In the evening they all came up to our camp and drew some more maps of the country between here and White and Copper River. All seem familiar with the ground and all agree in essential points. The chief variation is in the number of days required to go from here to Nicolai's. That varies from ten to thirty. I collected and pressed flowers growing around camp. [Camp 31–4]

Wednesday, July 1. I went down to the Indians' camp and took a couple of pictures. The boys have not yet returned from up river. The sun is very hot today with heavy thunder showers westward. The boys returned about 7:00 P.M., loaded with partially dried moose meat, and bringing two skookum young bucks along. One more is coming. We expect another consignment from up the Pelly and then hope to get away.

Sam is very ambitious to learn to read English. He has a first reader which he spends much time on. Most of the subjects described are foreign to his ken. I have printed a lot of short sentences using only the names of common objects with which he is acquainted. It is his delight to read these over to a group of wondering natives.

I made a trial this evening to collect Indian words, slow work partly on account of Sam's poor interpreting. The language is jawbreaking. The best man I have found yet is the Irish-looking gentleman, Mr. O'Rafferty. He has an intelligent looking squaw and a large family of which he is very proud. She is constantly "mucking" [nursing] her children. I changed plates in plate holders in Harper's house. [Camp 31–5]

Thursday, July 2. Heavy rain this morning before we were up, and the sun is now shining hot. The river is falling about two inches per day. A great firing of guns announced the appearance of a raft from Pelly River which landed at the upper camp. On it were a man and two boys, a bad lot. No packers among them. The Indians at first expressed great willingness to go to Copper River with us but now are hanging back. They have run across some superstition which deters them. The man that strikes the copper will die, so Sam says. [Camp 31–6]

Friday, July 3. At 9:30 A.M. I started for the mountain on the south bank of the river, northwest of Fort Selkirk; crossed a low river terrace of sand and gravel to the spur of the mountain and up a gentle slope to the top. The rock exposure on the spur is a reddish-grey gneissoid hornblende granitic. It seems to underlie thin layers of black vesicular basalt. The summit of the mountain is 1,030–1,100 feet above the river. Exposed rock is a greenish-black compact basalt with large porphyritic crystals, probably augite? weathered out of matrix leaving impressions. This mountain is probably the volcanic neck from which the basalt mesa between the Pelly and Yukon was extruded. It has been cut through by the river and the south side gouged out, leaving a steep slope. There are high gravel terraces up river about the level of the basalt mesa, that probably formed when the river was dammed by the lava flow.

Far to the south, thirty to fifty miles, are rather high mountains with some snow in patches, probably not permanent. The country between and to the southwest and west is rolling, with no land higher that one thousand to twelve hundred feet above the river. There is a heavy thunder storm to the north.

A family of Indians arrived while I was gone, no packers among them. We had a big powwow this evening. By much persuasion and many promises we have got five boys who halfway agree to pack for us. We will cut our outfit down to the lowest point and start with those if they can be started. We offered them $1.50 per day in silver on return for the trip. [Camp 31–7]

Saturday, July 4. The situation remains in status quo. Five Indians agreed to go but Schwatka backs out, hunts for excuses and obstacles. He makes going conditional on having two extra packers to go for a week or ten days and then return. We have a few warm words. The next plan is to take a wooden boat—one built last spring by Russell on Lake Aklene, now lying on the bank—and go down the Yukon and try going up White River. This appears to be the next best plan.

We have a report of copper being found one day's march across the river to the north, and we attempt to get Jackson to go and bring some.

He agrees but afterwards backs out and returns the money given to clinch the bargain.

The Indians celebrate the Glorious Fourth by a great deal of shooting, and blowing up a wooden cannon. Some have seen the miners of Fortymile Creek celebrate. The Indians have all moved down to the lower camp, packing most of their outfit on dogs. [Camp 31–8]

Sunday, July 5. Schwatka is awful glum, has nothing to say to me. We get at the boat and spend the forenoon repairing, pitching and nailing up. About twelve noon the Indians report that white men in a boat are coming up stream. Ande Bowker and two Indians from Forty-mile come from Harper, going across to look for copper. We invite Bowker to eat dinner with us and afterwards have a long talk about trails, copper, etc. [Camp 31–9, 10, 11]

Wednesday, July 8. We have spent the last three days making preparation for the portage and expect to start tomorrow. Our party has partly combined with Bowker. We have a promise of five Indians, four packers and Jackson, who promises to act as guide, and seven dogs. We had the dogs up last night and tonight to feed them; ravenous wild beasts. They eat an enormous amount of fresh and dried meat. One ate three-fourths of a cheese last night. He looks happy tonight. This afternoon I took six pictures around here and packed up instruments for shipment to Washington.

Eleven o'clock and it is too light to sleep well. I saw a star, the first since coming north. For a couple of days it has been hot in the middle of the day. Monday night was too sultry to sleep until about twelve midnight. I paraded the river bank to get cool. The boys go in swimming every day. [Camp 31–12]

Thursday, July 9. We spent the forenoon getting Indians and dogs together and making up packs. We have five Indians and seven dogs. I counted the Indians, not including three from down river: nine men, twelve women, nine boys, four girls, three babies.

We started at twelve noon, the sun very hot, and had a fairly good trail all day.

(Gul-In-Ti) (Tschut-Sai-A) = Jackson's dog

We went into camp in a scattering of spruce, on the side of a valley that is mostly open, with moss and moose brush, at 8:30 P.M. The 'skeets are rather bad. [Camp 32]

Friday, July 10. It is a hot morning, the Indians kicking. They are not satisfied with the grub and talk of going back. The difficulty is caused by Jackson, who wants to eat with us, as Bowker's Indian eats with him. We gave the Indians some flour and lard and they are all right.

We marched ten miles, following valleys that are mostly open with grass, moss and moose brush, or sparsely wooded. We followed the main creek to Camp 33, the last three miles hummocky and hard walking. The dogs got very tired, with disconsolate howling on starting after a rest. We found ripe strawberries on the side of a hill, and moss berries and blue berries are getting ripe. After going into camp I went downstream and caught a mess (five) of trout. The mosquitoes are very bad. They are small and come through the net. [Camp 33]

Saturday, July 11. We started at 9:15 A.M., leaving the main stream and taking a side valley leading to the southwest. There is no trail and it is very hard walking through the deep moss and grass hummocks. On climbing up a ridge I found exposed conglomerate, highly-changed mica schist with flattened pebbles. At a high point between two branches of the creek, 25 mi.+400 was black hornblende schist. West of the valley in which we lunched, black compact basalt was exposed. We found a lone Indian grave with arrows placed on it, scooped out. We climbed to the main divide, a long ridge covered with moose brush and moss, no timber. It is rounded, with a few exposures of basalt. The mosquitoes are very thick. We had a good view to the southwest across the valley of Selwyn River to the high mountains forming the divide between it and White River. The general course to the top of the next range is S60°W. There are some patches of snow beyond Selwyn River. We found a good game trail down the spur to the valley, crossed the east fork through a level swampy tract and reached the main branch of the Selwyn, went up a short distance and camped on a gravel bar.

The east branch is the smaller of the two, eight feet by three inches(?) on riffles. The west branch, sixteen feet by three inches on riffles, evidently at low stage. There are many gravel bars and the creek breaks up into many channels. I took a couple of pictures before lunch from

24 mi.+21 looking back on the trail and forward including the pack train. After supper we took a lesson of Bowker in gold panning near camp. We took gravel from the top of decomposed-granite bedrock and got one small color. [Camp 34]

Sunday, July 12. The trail led up the creek and we took the east fork, in a narrow valley. Snow lay in patches. We climbed up to a high rounded divide between forks of Selwyn River. Smooth outlines are formed by moss-covered slopes composed of masses of angular granitic fragments covered with a thick mantle of moss, dwarf willows, grass, etc. Through this many sharp points and pinnacles project. There is no sign of glaciation, no traveled boulders. I took a couple of pictures of the valley at Camp 34 and of rock pinnacles on the divide.

David shot a caribou in the afternoon. We went down to a scattering of timber on a branch of the Selwyn River and camped on the tundra. Caught a mess of five grayling after supper. Mosquitoes have been worse today than at any time on the trip, especially toward evening. The dogs had a good feed and can curl their tails once more. [Camp 35]

Monday, July 13. We had heavy walking this morning, on hummocks and moss. The first ridge after leaving Camp 35 is of schistose quartzite, somewhat gneissoid, then black basalt, generally vesicular, which continues to Camp 36. The Selwyn-White River divide is not so high as the one we passed yesterday, and held scattered spruce timber to the top. The country shows no trace of glaciation. There are sharp castellated rocks and pinnacles, and no rounded gravel or transported boulders. We camped at the forks of the creek. The country here is much wetter than about Fort Selkirk, and we have seen heavy thunder showers every day since starting. [Camp 36]

Tuesday, July 14. We followed a small stream nearly west all day; the valley is one-fourth to three-fourths miles wide, with scattered timber, moss and hummocks. Several branches come in from the north from fair-sized valleys. We crossed the main stream below its junction with the branch we have been following and camped. The valley is one and one-half miles wide, extending south.

Klo-Tas-Sin-Dik = Grass creek [Klotassin River]

The water is very clear, slightly amber in color, about twenty-five feet by fourteen inches on riffles, evidently at low stage. There are rather large gravel bars and the gravel is fairly well rounded. In the branch followed by the trail the gravel is very angular with few kinds of rock; no transported boulders. We went out after supper and got a mess of trout. [Camp 37]

Wednesday, July 15. At 5:30 P.M. we reached a high divide, the surface a porphyritic gravel. We had a fine view of the great expanse of White River with its many windings.

There is a large lake [Wellesley Lake] on the west side of the valley, crescentic-concave westward. The creek valley heading is about S60°W.

There are many fires in the valley, obscuring the view, making the high white mountains S45°W, very indistinct.

Within a radius of thirty miles the surface is very much cut up. There are many peaks and ridges about the same height as this divide, none more than seven hundred to one thousand feet higher. The streams have cut deep V-shaped valleys; no cañons. There are generally smooth green slopes with sharp castellated pinnacles of rock projecting through. The rock is porphyritic granite, much decomposed. At Camp 38, a cliff twenty feet high formed by cutting of the creek is so soft that the granite can be dug out with a stick. Bowker shot a caribou at 76 mi.+310. We camped on the side of a bluff in the first spruce timber. [Camp 38]

Thursday, July 16. A hot day, and I had a hard headache all day. We marched down valley through small spruce, the last three miles across an open valley with tundra and small ponds, many huckleberries and moss berries. The moss is frozen at a depth of about a foot. We crossed the creek which we have been following, twenty feet by six inches. A couple of miles above our crossing the creek forks, the two valleys about the same size. After crossing, Jack fired a gun and soon we heard a return shot. The Indians were all excited. We reached the river [Nisling River] about 7:00 P.M., and found Indians camped, waiting for salmon. Two families are on the east bank and four on the west. Those on the east side have just returned from hunting, loaded with dried moose meat. We camped in a spruce grove on the east side of the river. After supper Russell and I went out to a point of bar and panned some gravel; gold very fine, fifteen to twenty small colors. No good according to Russell. [Camp 39–1]

Friday, July 17. We decided to stay over a day and recuperate. The Indians are busy putting in a salmon trap. The trap is a fence across the river made of wattles tied to stakes. At a point down stream are conical baskets of slats. Fish going up stream strike the fence, and following it to get around, go into the open end of the basket and as it grows smaller can't turn around, and can't back out on account of the current.

We washed up clothes and loafed all day. Picked a mess of wild currants, about like cultivated but somewhat elongated. An Indian here has been to Nicolai's. Last night he was willing to go over with us. Jack has been scaring him by telling him he will not get paid and now he won't go. A brother-in-law of Harper, Alfred, is here. He was sent by Harper for copper and got only this far. He is to go with Bowker, and is a fairly good interpreter. There was a big powwow after supper, including a dramatic scene around the camp fire with Jack debating on the difficulties and dangers of the pass. It looks as though he might scare the others out. He poses for a big medicine man and spends most of the night howling and chanting, making medicine. [Camp 39–2]

Saturday, July 18. We were slow getting started this morning; we will have to wade the river. We bought about fifty pounds of dried moose meat from the Indians for tea and tobacco. Some is very rank but the best is not bad eating. Five salmon were taken from the trap this morning. Bowker joined our mess, the Indians will be put on too. We have meat with tea and a little flour.

We crossed the river at 3:30 A.M. this morning—one hundred twenty-five feet by three feet—on a rather swift riffle below the fish trap. There are large gravel bars with considerable variety of rock on them and well-rounded gravel.

Expedition stock has been up and down. Old Jack is playing a game. At one time he refused to go and also Sam and Denis said they were going to stay at the salmon trap. Finally we secured another boy as a packer and then an Indian who knows the location of the copper. This broke up the strike and Jack and the boys decided to go.

We started at 3:45 with seven packers and three dogs. Bowker has three packers and no dogs. We traveled four and one-half miles across the level valley, much drier than east of the river, through natural meadows with grass shoulder high, some patches of moose brush and willow, and some hummocks. Most of the valley has been burned over, burning off about a foot of turf down to coarse, well-rounded gravel. No glaciation of the country hereabouts. White River evidently heads in a

moraine lake and this is an old valley deeply filled with glacial gravel.
We crossed a low ridge to another wide valley with a small stream
flowing northwest, and camped on a small lake. [Camp 40]

Sunday, July 19. We made a good drive of eight miles this morning and
found very little water on the trail.

We lunched high up on the side of a mountain and kept well to the
divide between streams flowing into White [Donjek] or Ripple [Nisling]
River. We left granite for eruptive rock; greenish-gray, serpentinized,
and black slate; compact porphyritic rock, black siliceous schist, and at
Camp 41 compact fine-grained, black basalt, with some porphyritic crystals of feldspar.

The last two miles were in a drizzling rain. We camped at a small
stream eight feet by six inches, and put up a fly. [Camp 41]

Monday, July 20. A steady drizzle this morning. We slept late (10:00
A.M.), spent the morning drying out, and with signs of the weather breaking decided to start. Got off at 2:15 P.M.

At 8:10 P.M. we saw a caribou on the mountain and stopped while
the boys went after it. They came back in a couple of hours with part of
the meat. We went into camp in the first large timber, a scattering of
spruce eighteen inches through at the base but not more than thirty
feet high. It continued to rain during the evening and part of night.
[Camp 42]

Tuesday, July 21. It is a rainy morning and we are slow getting started.
The Indians don't want to go for the rest of the caribou they killed last
night. We started about 10:00 A.M. and got down to a creek flowing
northeast. Much gravel is evidently coming from a glaciated region but
there is no sign of glaciation here. We went up a branch to the south
and passed from diabase(?) with porphyritic crystals to porphyritic granite.
A heavy hail storm after noon. At 126 miles the valley has steep sides
and an irregular, hummocky bottom composed of white sand and some
boulders. Further south are three lakes, one a half-mile long, formed by
moraine dams. This is the furthest south any sign of glaciation has been
seen. There must have been a small glacier in the bottom of the valley
as the granite on the sides of the mountain stands up in sharp pinnacles.

We left the valley across a flat divide composed of sand with boulders, and from the south side had a view of the wide valley with steep mountains beyond. Far to the south (forty miles) are sharp snowy peaks, probably the St. Elias Alps. The valley is filled with boulders, sandy along the center. Bluffs along the small creek are fifty feet high, with slides in many places.

At 8:30 P.M. we reached the junction of the small creek draining the valley to the east and the large creek [Kluane River] flowing through the narrow valley from the south. We made 19.2 miles today. The stream from the east is clear, the stream from the south white with glacial mud.

After supper three Indians came into camp. They have seen us all afternoon but were afraid to come in. They are camped up river. [Camp 43–1]

Wednesday, July 22. We got up at 7:30 A.M., the strange Indians still in camp. We sent them with two boys up river for rafts and dried meat, then waited until late in the afternoon when four rafts with about a dozen Indians appeared and pulled into the mouth of a little creek. We tried to start Albert and David to trade but Jack stopped them, playing his old game. He must get something out of everything going. He is laying up wrath for a coming day.

There is no chance of our getting away today and we decide to start in the morning, without Jackson and his gang if necessary. The boys traded for dried meat and Jack set its price. After supper the Indians came around for medicine. We prescribed Vaseline, Frank Millay Leather Preservative, and Cuticura; got lots of meat in fees. [Camp 43–2]

Thursday, July 23. We spent the forenoon fixing up rafts and dickering with the Indians. The old Chief agreed to go with us to the copper but backed out at the last minute. We got started at 2:00 P.M. on three rafts; Schwatka, Russell, and I on the last with three Indians. The Indians had told us of a bad cañon a short distance down river; it proved to be merely a riffle, and we had no difficulty in running it. The name of the river is the "Klu-An-Tu," [Kluane River] according to Jack.

We ran down ten miles, the width averaging about one hundred yards, and were not able to touch bottom with a twelve-foot oar. It is very full of white sediment. The banks are of white sand with stratified layers of peat, with some bars of coarse gravel. At the mouth we came into the main branch(?) of White River, the "Don-Jek-Tu" according to Jack [Donjek River], quite different in appearance from the Kluantu. It is spread out with many islands and bars, very much more mud, and carrying a full load of fine white sediment. The banks are of quicksand, and we can drive a long pole out of sight.

We floated down White River four miles and landed on the west bank. On seeing smoke across the river the Indians fired guns and soon a crowd of nine Indians appeared on the opposite bank. After a great deal of shouting and a delay of an hour we marched down river a mile and camped on the bank. The Indians want to wait for a man to cross from the other side who knows the trail to the copper. Three of the boys crossed the river after supper. [Camp 44]

Friday, July 24. This morning we waited for the boys to return from over the river. They came across but landed on an island and appear to be unable to get over.

We have seen many fresh tracks of moose, caribou, and bear on the sandy river bank. This morning we saw a caribou cow and calf coming up the bank toward camp. The Indians all rushed out with guns and dogs and fired a volley but failed to get hits, excitable as children. The boys got over with the strange Indians and we talk, but we can't get one to go with us. We traded for dried meat and bladders of moose grease. The Indians demanded medicine as usual and we prescribed Cuticura, as the Vaseline is getting low. We succeed in getting the Indians started at 11:20 A.M., went half a mile and heard the dogs barking. Bowker and Albert started off and soon found a wounded caribou brought to bay by our extra dog. They brought him down with three shots, cut up and took some of the best meat, and gave the hide and carcass to the Indians from over the river. This delayed us over an hour and we got started again at 12:30. The walking was bad, through thick brush, dead timber and moss, and the day very hot, like August in the U.S. We camped between two creeks after a thirteen-mile march. [Camp 45]

Friday, July 25. We crossed the creek over very heavy boulders and coarse gravel bars, then turned up a gap and ascended 250 feet into a basin deeply filled with boulders and gravel drift. On the surface a deep layer of white volcanic sand formed steep alluvial cones at the base of the steepest slopes.

At 7:30 P.M. we came over a divide and opened up a grand view of the many snowy peaks of the St. Elias Alps. One large glacier came down to the valley, the highest peaks hidden by clouds. We were not able to identify St. Elias. I took a couple of exposures of the the range although it was rather late. We followed on to camp two miles further at the first pine trees, scrubby and trailing on the ground. Before supper the Indians wanted to go over the next ridge to timber but after filling up with grub they want to camp. [Camp 46]

Sunday, July 26. I went up on the ridge ahead of the packers and got some topography. The mountains were obscured by clouds but I had a good view of the valley and glacier. We had good walking, the moss not deep because of the dry climate, and had lunch at a clear stream, thirty feet by twelve inches, that probably drains the lake up valley. After lunch we got on to rough ground; small lakes and potholes with gravel sides.

About 3:30 P.M. we reached the bank of the creek and found high water; a roaring yellow torrent, with a bed of big boulders that we were not able to cross. We went downstream one-half mile and camped to wait for the water to fall. Rain began and it rained until night. Albert and David went downstream several miles, and reported low ground and the river broken up into numerous channels. [Camp 47]

Monday, July 27. A bright morning. The Indians were glum and slow to pack up. The river has fallen about a foot during the night. We started down stream at 9:35 A.M., went about two miles and waded across. There are two principal channels, the water three feet deep and very swift, with bars two-fifths of a mile wide and numerous small channels. The Indians put down their packs and wanted to leave them and go on to the copper without them. The difficulty originated with the big buck from the fishing village. Things looked blue for a few minutes but we made a potlatch of a dollar to each and they took up the packs again. We turned south and marched two and one-half miles through open, dry, park-like country, crossed a small steam of milky water and stopped at 1:30 P.M. for lunch. I have seen many fragments of pumice as big as a fist. On the bluff west of the river, there is white tufa on top of gravel, probably one hundred feet thick.

After lunch we continued south for two miles through open park-like country and struck a moraine composed mostly of white tufa with many angular fragments, up to boulders three feet in diameter of amygda-loidal lava, conglomeral breccia and some granite. We climbed about one hundred feet to the surface which is very rough and full of pot holes, with ice seen in a few places. The surface is partly covered with moss, huckleberry bushes and some small trees.

We went down to the river bed and found it to be fifty yards across, very swift and probably deep. We were unable to cross so returned on the trail and crossed on the glacier above the tunnel from which the river comes. The Indians were badly scared. Jack got to a bad-looking place and made medicine. I went over first and the rest followed. Very little ice shows at the surface which is covered with mud and stones, and crossing [Klutlan Glacier] was not dangerous. The Indians were greatly relived when across. They have a strange superstition; Jack told us at lunch not to fry any grease or the ice would break in. The Indians all assured us that after we had seen the ice we would not think of going on to Nicolai's.

We camped at a small stream near its mouth, a torrent over a bed of boulders. We will have to bridge it in the morning. [Camp 48]

Tuesday, July 28. Einstein built a bridge before breakfast. We crossed open country with hias [many] huckleberries, but very few trees, some along the creeks. It is a rolling plain of white drifting volcanic sand, with some moss and bushes trying to grow in the sand. There were some clear lakes and small clear streams full of fish, and we crossed several muddy streams flowing from the mountains in gravely channels. We stopped about 4:00 P.M. and the Indians started out to hunt. We had been sitting down an hour or so when we saw a group of caribou on the plain across the creek. The Indians surrounded them and we saw an exciting hunt, with a great deal of shooting, but not much game. Einstein shot one with the shot gun and Nee-Du with the Winchester. Later David came in with one and now meat is plentiful in camp. [Camp 49-1]

Wednesday, July 29. The Indians began stirring early. After breakfast they started up the creek [Kletsan Creek] toward the mountains looking for copper. We soon followed and found them scratching in the gravel, finding many small copper nuggets, one-half to one inch in diameter. We went on up the gulch looking for the ledge but found no encourag-ing signs. I climbed a point between two streams, finding dark greenish-black volcanic rocks and some red sandstone or jasper. On the steep talus slopes the rock breaks into fine fragments. Limestone at the lower end of the gulch was folded in with bedded eruptive rocks. The limestone contains many fossils, mostly crinoids and large fine corals, probably Devonian?

I got back to camp about 7:00 P.M., all the others in before me. Bowker and the Indians are getting ready to start back. Schwatka settled up with the packers, about $25.00 apiece. They take a stirrup cup of quinine and leave. [Camp 49–2]

Thursday, July 30. We packed up and got started about 8:30 A.M., our belongings cut down to the lowest possible notch. Each has thirty-five to forty pounds of grub and blankets, etc., making packs of seventy-five to eighty pounds. We stopped at 3:30 P.M. as we didn't want to get too tired the first day. We marched eight miles today in a slow drizzling rain and camped in spruce timber at a small brook. It is hard work but we will make it. [Camp 50]

Friday, July 31. It rained hard in the night, getting our blankets wet. We got away at 8:30 A.M. and crossed two creeks by wading through cold water. The second comes from a glacier. The wind came up and it was so cloudy that we got no good view. The glacier [Griffin Glacier] is probably a mile across and very level on top with an abrupt front. The river spreads out about a mile with many channels. We camped on the west bank. [Camp 51]

Saturday, August 1. It rained hard all night. We built a wickyup but it didn't turn the water and we were flooded out at 3:00 A.M. We got up, made a cup of tea and tried to dry things out, but no good. We had breakfast and got off at 6:00 A.M., still raining. The packs are very heavy. We marched through spruce woods with much standing water and got very wet. The river from the glacier rose a couple of feet in the night. We crossed two small streams, very high, filled with yellowish-white mud. The white volcanic tufa has wholly disappeared. The wash in the streams is angular, granite and black slate.

At 9:30 A.M. we stopped under a big spruce to build a fire; started again at 1:15 P.M. We marched all day through timber, moderately thick with much brush. The surface is generally even and sloping gradually northward for the first two miles after leaving Camp 51; rough with many large boulders of dark volcanic rock. Our route lay midway between the river to the north and the mountains to the south, about one and one-half miles from each. It has rained all day but with signs of clearing. The mountains have for the most part been entirely concealed, but the clouds lifting gave us a few glimpses. They present a very steep, abrupt front to the valley. The sides facing the valley are covered with snow and small glaciers from which come the many small muddy streams we crossed. [Camp 52]

Sunday, August 2. It rained hard in the night but we were fixed for it and kept dry; clearing weather this morning. We left the timber and marched on the broad level gravel plain bordering the river. There was no irregularity more than two feet high, the surface of compact sand and gravel, a few pebbles six to eight inches in diameter, much black slate and sand, quartzite in gravel with much volcanic rock. There were only a few tufts of vegetation, the river probably overflowing every year. It is about three miles across to the bluffs on the opposite side of Nau-Sin-Klat River. From the bluffs (fifty to seventy-five feet) is a gradual slope of a mile to the foot of the hills, green to their tops, probably three thousand to thirty-five hundred feet above the valley. Small streams coming from the mountains to the south form symmetrical alluvial cones where they flow from the timbered terrace to the gravel plain. In motion of cutting a channel in the terrace they build up a cone on the lower plain.

We turned southwest. River channels occupy the whole valley from one mountain to the other forcing much wading. A big current cuts the bluff on the east side and we have to turn back and climb the bluff. We camped in the last timber on the east side. The valley is occupied by a glacier about a mile above, with moraine around, and several smaller glaciers come in from the sides. The river issues from the extreme west corner. Clouds fill the pass so we can't see what is before us. We cooked up enough grub for two days. A strong wind is blowing down from the glacier. A thin seam of volcanic ash shows on the opposite side of the river and some in favorable places on this side. [Camp 53]

Monday, August 3. The wind blew all night, the first night since leaving the mouth of Taku River that the mosquito bar has not been needed. We got started at 7:30 A.M. and got wet crossing the many branches of the creek coming from a narrow gorge to the east and from the east side of the glacier. We climbed the steep face of the glacier, which has four to six feet of moraine material on the surface, with ice showing in places. The moraine surface is very rough and is covered with vegetation which decreases and disappears in a couple of miles. Very hard walking. Cones and ridges of ice one hundred feet high are covered with a thin layer of stones. On reaching a small tongue of white ice toward the west side we had fine walking for a couple of miles, but it soon ran out and we got onto the moraine again.

We reached the summit about 4:00 P.M. and were much surprised to see water, a river, west toward a narrow gap. The glacier has a smooth regular surface in this direction with little moraine on the surface and but a small terminal moraine. We made camp on the north side of the stream one-fourth mile below the foot of the glacier, with enough willows for a fire.

There is a very high peak, partly exposed by drifting clouds, bearing S32°W (var. 30°E). The big glacier [Russell Glacier] which we have crossed comes from this, which may possibly be St. Elias. The mountains on both sides of the pass in which we are camped are composed of horizontally-bedded rocks. At camp the rock is amygdaloidal basalt, some rocks of bright red sandstone; vertical dikes. [Camp 54]

Tuesday, August 4. We made a good march of seven miles in the forenoon along the north side of the river by a deep cañon of basaltic rocks, and stopped for lunch on the edge of a sand plain in front of the glacier [Frederika Glacier] coming from the north. While getting lunch we saw a bear a little way off eating berries. Russell took the rifle and crawled up, with no cover, but the bear winded him. He got a shot but missed, and the bear ran. He fired a couple of more shots but didn't hit, and the bear went up the side of the mountain at gallop.

We started at 2:00 P.M. after taking a picture of the glacier front. The stream from the glacier was deep and swift with much ice running, and we were unable to wade. We went up to the glacier but could find no passage, so returned and made camp in the willows hoping the stream would fall in the night. [Camp 55]

Wednesday, August 5. Rain began in the middle of the night and kept up steadily all day. We tried to wade the stream from the glacier but found it too deep; the rain prevented its falling. We came back and near camp finally crossed the main stream, a deep wade but we made it by holding together. It was hard walking over the moraine of the glacier on the south side of the river. The rain has made the surface a sliding mud quicksand. We crossed the second glacier moraine, finding it not so bad as the first.

There is a great quantity of fossiliferous marble in the moraine, resembling Tennessee marble; mostly crinoid stems, and a few brachiopods, some red and pink. The river flows in a narrow rocky cañon, and we had to climb high up to avoid gorges. The brush is very bad.

We stopped for lunch in the first spruce stand, with large trees from fourteen to sixteen inches in diameter. After lunch we went down to the river bench but were drawn up by the current cutting the bank. We followed along a talus slope below a high cliff of limestone, nearly horizontal. This limestone is interbedded with quartzite, conglomerate, sandstone and trap sheets. The topography is characteristic of horizontal rocks with Grand-Cañon type towers, pinnacles, cliffs and flat-topped mesas. We made camp about 350 feet above a lake into which the river flows with a broad delta. There are many icebergs floating in the lake.

After supper I went ahead on the trail to Nicolai to see the country. A large glacier [Nizina] coming from the north has pushed its moraine across the valley and dammed the river, forming the lake. Three glaciers unite at the foot and are marked by heavy lateral moraines. The one on the east leaves via the high mountains north of the river we have been following and probably is continuous with the glacier whose stream turned us back. It does not reach the lake but its moraine does, with the ice back a mile or so. The middle glacier heads back of a sharp mountain. In the center of the valley is a needle peak, with horizontal beds curving around its base, concave to the west. It carries a number of well-marked medial moraines where first seen. The western [Nizina] glacier, largest of three, comes from a high peak [Regal Mountain] probably one of the Wrangell group, its whole course in sight. It covers the whole side of the mountain like an enormous cascade, with no medial moraines. [Camp 56]

Thursday, August 6. It rained some during the night but was clear all day. We started at 8:30 A.M., following a bear trail. I took a couple of pictures of glaciers from a knob. The brush on the mountain side is almost impossible; we got along by following the bear trail. On passing the point where the river is dammed I noted that the glacier [Nizina Glacier] has pushed across the valley and the river now flows in a narrow, rocky channel, then plunges in a hole and disappears under the moraine-covered ice, reappearing a half mile further down. There are many fine ice cliffs 100 to 150 feet high along the river below. We crossed a smoothly-glaciated mountain point of limestone, got down onto a bench at the lower end of the glacier and had good walking to our camp in spruce timber at the base of a cliff 800 feet high. [Camp 57]

Friday, August 7. We started down from the bench on the west side of the river. The current strikes the bluff perpendicularly for 1,000 feet and we could find no passage so were compelled to cross, with some tall wading—fair walking on the west side. The creek [West Fork] comes in from the west, the deepest channel yet crossed. We were soon driven from the bench to the timber where we found bad walking with boulders from the cliff, fallen trees and brush. The last two and one-half miles of the day's march were the worst we have had; mostly on the side of the bluff with talus slopes, black alder, slides, and raining all day.

We made camp after dark on a flat point in spruce timber. [Camp 58–1]

Saturday, August 8. It rained hard all night but we kept comparatively dry. I crossed the point after breakfast. The walking looks bad and there was no sign of Nicolai. We decided to build a boat. Our grub is getting very low. [Camp 58–2, 3]

Monday, August 10. We have just finished our boat. It has a frame of poles tied together, is covered with canvas of Schwatka's and my bed covers and is pitched with spruce gum.

The river here flows in a cañon about two miles wide, between limestone bluffs one thousand to two thousand feet high. On the west side at 139 miles is an over-thrust fault. The fault plane dips S20°W. Heavy limestone beds slide over each other while some beds of black shale are highly contactic. This is the first marked disturbance in bedding noted since coming over the divide. No fossils were seen in this limestone; it contained some nodules of chert. A larger block of black slate on the beach was filled with brachiopods, probably Triassic?

About four miles below camp the river turns sharply to the west, flowing between bluff banks 150 feet high.

The country to the south is level or rolling for eight to twelve miles and then a steep mountain range. Sharp mountains lie to the north. Along the river white contactic schists? with black dikes? are exposed, and there are horizontal black shales and gravel capped by probably more recent gravel.

After running about twenty miles we stopped for lunch at 2:30 P.M. After lunch Russell went ahead to look at the river, which makes a sharp turn to the north apparently entering a cañon; Schwatka went to the bluff to set signal fires.

We had come about six miles this morning when we struck a gravel bar and snagged a hole in the bottom of the boat in trying to land. Russell jumped out and got wet. We held up and put on a patch. We have struck several bars since but with no serious damage. The river is very swift and swings back and forth from bluff to bluff, breaking up into many channels and making much paddling necessary to keep in the channel.

At 5:00 P.M. we passed a good-sized stream coming from the northeast. It has a rather wide valley and looking up, very high snowy peaks appear. At 5:15 P.M. we snagged another hole in the boat and stopped to patch. We have passed some very rough water, shipping some over the bow. After passing a low point we camped on a sand bar in front of a bad-looking hole which we want to inspect before running. [Camp 59]

Tuesday, August 11. We pulled and waded across to the north side of the river, and Russell and I climbed the bluff to inspect the river. Ahead lay a bad cañon, very crooked, with sharp points jutting out from the sides, on which the water strikes. The course of the river turns southwest for four or five miles. We came back to consult; we considered abandoning the boat, but finally decided to take our chances and run the river. The next five or six miles were very bad; very rapid water continually dashing against rocks on one side or the other. We made several landings and climbed up to inspect. Finally about 1:30 P.M. we emerged from the cañon at the junction with the Chittyna [Chitina River] and stopped for lunch. The Chittyna appears to be the smaller stream and with clearer water, but flows in a larger valley. Probably larger at some season than the Chyttystone [Chitistone].

We left the junction at 1:35 P.M. The current is swift, probably averaging seven miles per hour. The river flows between two hundred-foot bluffs of horizontally-bedded sand, silt and gravel, about one to one and one-half miles between bluffs. The current swings back and forth and there are many bars.

We ran until 5:30 P.M., the current decreasing somewhat, and went into camp—Number 60—at the foot of a gravel bluff at the sharp bend shown on Allen's map. A strong head wind was the cause of our camping. We fried a mess of rose berries with cranberries for supper. It was not a success and made Russell sick. We boiled the last mess of beans tonight. [Camp 60]

Wednesday, August 12. We got off at 6:25 A.M. this morning. The river has risen over two feet during the night. The country continues about the same as yesterday but the bluffs are more rocky. At 9:00 A.M. we came to the junction with Copper River and about two miles below our sight was gladdened by Taral. There are two Indian houses, no sign of Indians.

We landed a short distance above and as we came down the shore Indians appeared. Nicolai, whose existence we had come to doubt, appeared in Boston clothes and gave us welcome. We were invited into his house and given a feast of bread, bacon and tea. Nicolai speaks some Chinook and English, and we understand him fairly well. He is about forty-five or fifty-five, rather slight but well formed. We learned that he is going down to the mouth of the river in a few days. A large skin boat is being made. He is waiting for the river to go down, and for the arrival of Indians from up the Copper River who are to accompany him. The boat is covered with seal skins. We got some flour and bacon from Nicolai and later about two-thirds of a sack of flour from another Indian and feel comfortable on the grub question.

Salmon are plentiful, both dried and fresh. The first run is about over and the Indians are catching only a few for present use. We got two big ones each day for a quarter. [Camp 61–1]

Thursday, August 13. Russell and I brought our boat up on shore and turned it up for protection against the strong wind, which continues up stream except for a short quiet time in the middle of the night. Schwatka sleeps in Nicolai's house.

We spent the day loafing. We are all tired and glad of an opportunity for a good rest. Nicolai says he will start down the river in five sleeps.

I went to the foot hills back of Taral. The rocks are silicious, talcose schist, with some quartz veins. Granite is abundant in the wash of a small stream coming from the east. The rock for ten miles up Chittyna River has been trap; greenish-black streaks of serpentine. [Camp 61–2]

Friday, August 14. We loafed around camp and ate salmon, and spent some time watching a volcano(?) to the north. I took a photograph northward. A boat load of Indians arrived from up Copper River. They are going down with Nicolai. They have a very decided hebraic cast of features. [Camp 61–3]

Saturday, August 15. By 4:20 A.M. I had hiked through thick moose brush, grass and moss to timber line on a bench of the mountain to the east and about 2,000 feet above Taral. To the east and west is a high plateau, with ravines cutting it into rounded hills with vegetation to their tops, 3,000 to 3,500 feet above the river, and a few rocky sharp peaks. From timberline to the river spruce is thick in patches but generally rather scattered with much birch and alder.

To the south are sharp, rugged, rocky mountains with no rounded outlines, 5,000 to 6,000 feet above the river, with some snow and a few small hanging glaciers, and some rather level bench land near the river after passing Woods Cañon. To the north is a broad valley, and between the Chittyna and Copper a gently-sloping bench 200 to 500 feet above the river, with high mountains behind. Beyond the first high mountain is a broad, smooth, snow-covered dome, and beyond it and slightly to the left is a sharp black peak. To the right of the peak a cloud of vapor constantly rises.

The broad white mountain is probably Blackburn and the sharp black peak is Wrangell, far beyond. The latter is volcanic, emitting steam which quickly dissipates. Other mountains to the left are concealed by clouds. The rocks at my point of observation are gray hornblende (granite?).

Nicolai says we will start for Eyak tomorrow if the wind doesn't blow. It is now blowing a gale. [Camp 61–4]

Sunday, August 16. We got up at 4:00 A.M. and after breakfast, loafed and waited for the noble Siwash; packed up and embarked at seven. We pulled across the river and left one Indian, landed again below and picked up "Blind Bart," and finally were started. At fourteen miles the first cañon looks bad; the river seems entering closer in, walls 100 to 150 feet high composed of highly contorted greenish talcose schist. The difficulty of running is not to compare with the cañon on Nizina— no high waves or bad whirls—the river is about sixty to seventy feet wide in one narrow place.

At 7:40 A.M. we landed below the cañon to shift the load and place boards under it.

We stopped for lunch at 10:30 A.M. on the delta of a small stream on the east side of the river, two miles below a stream coming in from the west. The mountains are steep and rugged and there is a glacier back in the ravine on the east side opposite the mouth of the stream. Another is two miles down on the west side, its foot about 500 feet above the river, and very steep. The rock is a hard dark-gray or bluish silicious slate or schist with some quartz veins.

At 3:30 P.M. we were on a delta-filled lake above the cañon. Two large glaciers put into this from the west. The northern is composed of three, coalescing at the base, four to five miles wide. The front is mostly covered with moraine and vegetation. At the north side ice shows but there is no cliff fronting the river.

At 4:20 P.M. the Indians think they see sheep on the mountainside. We go into camp while they go up for sheep. They don't find any. Camp 62 is at the foot of a steep slope. No timber, but covered with a dense thicket of black alder and devil's club. The alder is twenty feet high. The rock is a silicious slate. [Camp 62]

Monday, August 17. Last night was very warm; it is clear this morning with no wind. We got up at 4:10 A.M. and started at 6:00 A.M.

By 9:00 A.M. we left the lower end of the lake on a small channel west of a rocky island; the rock highly inclined and contorted, curly, gray silicious shales. The lake is formed by the moraine of a larger glacier coming in from the west. The moraine is wooded; ice shows at the north corner and slopes down to a sand plain.

The moraine along the north side of Miles Glacier pushes across the valley and crowds the river against the mountain. A small glacier pushes its moraine down from the west side opposite. Between the two are rapids. The moraine of Miles Glacier is composed of large angular blocks of gneiss, granite and contorted black schist. The gneiss contains fragments of black schist.

After passing the northern lateral moraine the river turns east and washes the front of Miles Glacier along an ice cliff 350 to 400 feet high. Bergs are constantly breaking off and the lake in front of the glacier is filled with floating ice.

We crossed the moraine and camped on a beach in front of the glacier. A bear was on the beach and Nicolai and Nahui went after it but failed to get it. It was crippled and took to the river. [Camp 63]

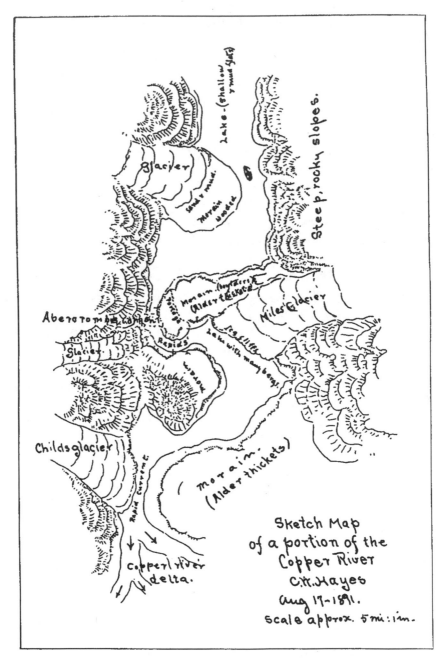

Map 4: Copy of a sketch map from Hayes' journal showing glacial conditions on a portion of the lower Copper River, August
17, 1891. Scale approximately 5 miles:1 inch. [See Hayes, Charles
W. 1898. "Copper River as a route to the Yukon basin." *American Geographical Society Bulletin*, Vol. 30, pp. 126–134, p. 131]

Tuesday, August 18. The Indians went back for the boat and got in at noon in a drizzling rain. They are afraid to cross in front of Miles Glacier, saying there is too much ice. Big bergs dropping off frequently make very rough water when they strike the lake. They decide to start at 3:10 P.M. We got ready and embarked in six minutes. We pulled past the front of the glacier and no ice fell down during our passage. The front, about four miles of ice cliff, recedes on the south corner. The river swings west and flows past the face of Childs Glacier with a rapid current. It is not deep and there is a good beach off the point. We ran past without difficulty and waited for Nicolai and Tannah Tyee who took the beach. We were not able to make Alaganik so camped on a mud flat with some brush. It rained hard all night. Russell and I slept in the boat. We have left the gravel bars behind and are now in mud-flat deposits. [Camp 64]

Wednesday, August 19. We started about 8:00 A.M., crossed to the west side of the delta and reached Alaganik about 11:00 A.M. There were two Indian houses, a miserable set. An Indian told us that the white men have all left Eyak and offered to take us to a cannery and trading post on the east side of the delta. We ate lunch and started back. I went with an Indian pilot and a boy in a dugout canoe.

We pulled up western channels five to seven miles against a strong current in pouring rain, crossed a broad center channel to one a couple of miles broad toward the east side of the delta and came down this one. We found the cannery about 4:00 P.M. on an island on the east side of the channel. We are all thoroughly wet and shaking. The cannery consists of several large buildings, all closed and the men taken away three days ago. There was no one to be found for some time; then we found Mike Duval[5] and Dick Temple who are in charge for the winter, and went up to Mike's house, got warmed and had a good supper. Two men, Cloudman and Boswell, are going to Eyak in a couple of days; we will go with them and try to catch the mail boat. We found that the Indian at Alaganik was posted by the trader, George Barrett, to turn Nicolai and his party this way and prevent them going to Eyak with their furs.

The channel we followed in the morning to Alaganik and back is the western-most of the delta channels. It is comparatively small, very winding but with good current. It flows rather close to rocky points of

5. Schwatka recorded the name "Duval" as "Duvall."

the mountains to the west. Clouds hung so low no good views were obtained. Some glaciers come down from this side but not large ones. The low land, and the slopes west of the channel, are wooded with good spruce timber. There is none growing on the delta now but along the western channel many large dead spruce trees and stumps are still standing below the present level of water, indicating a recent sinking of the land surface. [Camp 65–1]

Thursday, August 20. Heavy rain all night and wind blowing a gale from the northwest; can't leave for Eyak today. We loaf and eat. [Camp 65–2]

Friday, August 21. Still blowing a gale with heavy rain. No leaving for Eyak today. [Camp 65–3]

Saturday, August 22. The wind has gone down and it is not raining much so we may get away. We leave in the rain with no wind about noon. The wind soon comes up and we sail, keeping inside of the bar against which a heavy surf is beating. North of us or on our right are many low grass-covered islands, between channels of Copper River. A heavy surf is rolling on the outer bar. The channel, two to five miles wide inside of the bar, is mostly dry at low water. The wind died out and we pulled the last two or three miles to the mouth of Eyak River (see U.S. Coast Survey Chart, North West Coast of North America, Sheet No. 3, Icy Bay to Semidi Islands) and got across the flats just in time to avoid being stuck. (Copper River delta as given on the above chart is entirely erroneous.) Eyak River empties into the western channel of Copper River. North of the mouth three to five miles are three glaciers, the two northern coalescing at the base and spreading out fan-shaped.

Note: Captain Alexander tells me he has seen very large spruce stumps twelve feet below high tide, in Copper River delta, mostly in the channel along the west side.

The current was too swift to go up river and we have to wait for the turn of the tide. We will start at 1:00 A.M. We got supper at the fishing house. Russell and I made our beds in the Columbia River boat while the rest went up to the house. [Camp 66]

Sunday, August 23. We started upstream about 1:00 A.M. with the current, in moonlight, shooting ducks along the way. The river is one hundred yards wide, very crooked with many levies and low shallow banks.

The current soon slacked, then turned, and we had a hard pull through the narrow channel to the lake, about five miles from the mouth. The lake is two to three miles long. We reached the fish houses at the upper end of the lake about six and walked down one-third mile on the tramway to the canneries across the isthmus. The superintendents were both gone. We got meals at the Alaska Commercial Company house, loafed and slept all day. Captain Humphrey came back and offered to take us to Nuchek in the steam launch, *Salmo*. We started about 6:00 P.M., reached Nuchek about 1:00 A.M. and found the mail steamer had left the day before.

Sunday, August 30. On board Steamer *Salmo*. Coming back from Nuchek last Monday I accepted the invitation of Captain Marshall to remain on board the *Salmo*. I have spent the past week cruising about the Sound from the cannery to fishing stations at Gravina and Fidalgo. The boat has brought in 12,615 silver salmon during the week. At one haul of the seine in Gravina Bay I saw 1,011 fish caught. There is new snow on the mountain tops about Gravina Bay today.

Thursday, September 3. I came ashore from the *Salmo* and put up in Captain Marshall's room at the cannery.

Friday, September 4. A clear day, the first since we struck the coast. I took some pictures of the cannery, etc.

Saturday, September 5. The steamer *Jeanie* left today for the cannery on Cooks Inlet.

Tuesday, September 8. I accepted an invitation to go down Eyak River in the launch *Beaver*. We started directly after dinner, made down-river and across the flats to the fishing house on Alaganik Channel of Copper River and back to the mouth of Eyak River on the high tide, nearly getting stuck on the flats. I got supper at the fishing scow and slept in the launch. We started back on the morning tide and got to the cannery for breakfast. I took three pictures at the fish house on Copper River flats, and one of the glacier to the northwest.

Thursday, September 10. After dinner I started with Captain Humphrey and Russell in a bidarka across the bay and to the south. We followed the eastern side of Hawkins Island and crossed through by a narrow tide channel. There were rather heavy swells on the west side. We met the launch *Goby* about half way up on west side of the island, went up to Burn Point, and spent the night in Charley Rosenberg's camp. A fine clear day and night.

Friday, September 11. I spent the forenoon helping Rosenberg drag seine. We caught 991 fish, for which he gets five cents apiece. Got home about 3:30 P.M. Rained.

Saturday, September 12. We started soon after breakfast in the *Goby* for Burn Point, fishing with dynamite. Captain Humphrey fired six shots—six sticks of Giant powder each—and got 23–6–18–0–0–130 salmon. Very few other fish were killed except some small white minnows two to three inches long on which the salmon are feeding. It rained and blew all day.

Sunday, September 13. The crew all have headaches from the effect of the dynamite explosions. A fine day.

Monday, September 14. Russell and I started on the stern-wheel steamer about 8:00 A.M. down river duck hunting, taking along an Indian canoe. We left the steamer at the mouth of Bowker's slew, went down and found ducks in the ponds scattered over the meadows. It was very difficult to get within shot as there was no cover but grass twelve to eighteen inches high. We started back up river against a swift current at 3:45 P.M. A heavy rain came up, then a hail storm with one distinct peal of thunder—the first I have heard on the coast. We got home at 6:15 P.M. with nine ducks.

Tuesday, September 15. Loafed and got over the effects of the hunt. It has been a bright day with showers, cool but not unpleasant.

Prince William Sound is very irregular in outline. Numerous bays extend northward, each a fiord with a glacial stream at the upper end, usually with a considerable delta. At the head of Cordova Bay, Burn Bay, Gravina, Fidalgo and probably others are fresh-water lakes. Water in the bays, except at their heads, is very deep even close to the shore. Between bays are peninsulas in the form of high, rugged mountain ridges. Between Gravina and Fidalgo is a peninsula of probably 5,000 to 6,000 feet, wooded up about 2,000 to 2,500 feet, and above that moss and bare rock. The rock I have seen is black shale and beds of hard brown sandstone, highly contorted. The shale is quite slaty, slickensided, with steep and variable dips. In general the strike is about north and south, but variable. I saw no fossils, but plant impressions if even present would probably have been destroyed. The sea has stood at its present level long enough to cut a slight terrace and sea cliff. Forty to fifty feet above the present level there is a sign of terrace erosion and some level land at this elevation. The mountains are wooded, or covered with grass, and moss is noticeable on the west side of Hawkins Island, and Hinchenbrook Island at Nuchek.

From the Sound a high range of snow-covered mountains lies to the north of the bare mountains forming the headlands between the bays and at their heads. This is the source of the glaciers at the heads of the fiords. Only one presents an ice cliff to the sea, that west of Fidalgo Bay.

APPENDIX
CAMPSITES OCCUPIED BY THE
NEW YORK LEDGER EXPEDITION, 1891,
AS RECORDED BY CHARLES WILLARD HAYES.[1]

CAMP	DATE	LOCATION NOTES
	5/25	Leave Juneau aboard Native canoe.
1	5/25	Fish House on Taku Inlet owned by Frank Murray and company. (Map A)
2	5/26	Taku River, camp on low island. (Map A)
3	5/27	Taku River, camp on slough. (Map A)
4	5/28	Taku River, below south fork (Inklin River). (Map A)
5	5/29	Taku River, camp about one mile above south fork (Inklin River). (Map A)
6	5/30	Taku River, camp on gravel bar. (Map A)
7	5/31–6/1	Camp two nights at fork between Nakina (Nahk-a-náhn) and Sloko Rivers (Canoe Landing). (Map A)
8	6/2	Mile 4.5, Nakina River.[2] at the "The Nubbins." (Map A)
9	6/3	Mile 8.0, Nakina River, Ah-kah-tée village. (Map A)
10	6/4-5	Mile 10.0, camp two nights at fork of Nakina and Silver Salmon Rivers. (Map A)

1. Compiled by the author using Hayes' journal, consisting of two hand-written volumes (on file at the office of the U.S. Geological Survey, Branch of Alaskan Geology, Anchorage, Alaska 99501).

2. Distances shown are from Nahk-a-náhn by Hayes' estimate.

Camp	Date	Location Notes
11	6/6	Mile 16, camp on mountainside above Silver Salmon River. (See p. 70.) (Map A)
12	6/7	Mile 22.5, camp at Trout Lake (Katina Lake). (Map A)
13	6/8	Mile 27.5, camp on knoll. (Map A)
14	6/9	Mile 32, camp on boggy meadow. (Map A)
15	6/10	Mile 38, camp on small ridge in burnt area. (Map A)
16	6/11	Mile 45, camp at stream crossing (Hurricane Creek). (Map A)
17	6/12	Mile 51, camp on small side stream. (Map A)
18	6/13	Mile 55, camp in Ptarmigan Pass. (Map A)
19	6/14	Mile 63, camp in swamp on flats near lakes. (Map A)
20	6/15	Mile 77, Camp at lower end of fourth lake. (Map A)
21	6/16-17	Mile 81, Camp two nights near south end of Ah'k-klain (Teslin Lake). (Map A)
22	6/18	Camp on point of peninsula, east side of (Teslin Lake). (Map A)
23	6/19	Camp on bluff, east side of (Teslin Lake), near mouth of Nisutlin River. (Maps A, B)
24	6/20	Camp on east side of (Teslin Lake), north of Nisutlin River. (Maps A, B)
25	6/21	Teslin River, about three miles below Teslin Lake.
26	6/22	Teslin River. (Map B)
27	6/23	Teslin River. (Map B)
28	6/24	Yukon River, at camp of Swingler and Harvie. (Maps B, C)
29	6/25	Yukon River, about one mile below Little Salmon River. (Map C)

Camp	Date	Location Notes
30	6/26	Yukon River, below Rink Rapids. (Map C)
31	6/27–7/8	Yukon River, camp twelve nights at Fort Selkirk. (Map C)
32	7/9	Mile 10.5, first day from Fort Selkirk.[3] (Map C)
33	7/10	Mile 21, camp on stream. (Map C)
34	7/11	Mile 32, camp on branch of Selwyn River. (Map C)
35	7/12	Mile 43, camp on second branch of Selwyn River. (Map C)
36	7/13	Mile 54, (Map C)
37	7/14	Mile 67, camp on Klo-tas-sin-dik (Klotassin River). (Map C)
38	7/15	Mile 79, camp on side of bluff in spruce timber. (Map C)
39	7/16–17	Mile 90, camp two nights in spruce grove by (Nisling) river. (Maps C, D)
40	7/18	Mile 99, camp on small lake. (Maps C, D)
41	7/19	Mile 110, camp on small stream. (Maps C, D)
42	7/20	Mile 116, camp in timber. (Maps C, D)
43	7/21–22	Mile 135, camp two nights on Kluantu (Kluane) River. (Maps C, D)
44	7/23	Float Kluantu (Kluane) River, camp on White (Donjek) River. (See page 165.) (Map D)
45	7/24	Mile 13, camp near Koidern River.[4] (Map D)
46	7/25	Mile 30, camp just beyond divide. (Map D)
47	7/26	Mile 38, camp by roaring torrent (Generc River). (Map D)
48	7/27	Mile 51, camp beyond torrent (Generc River). (Map D)

3. Distances shown are from Fort Selkirk by Hayes' estimate.
4. Distances shown are from Camp 44 by Hayes' estimate.

CAMP	DATE	LOCATION NOTES
49	7/28-29	Mile 62, camp two nights on stream (Kletsan Creek) below copper deposit. (See page 171.) (Map D)
50	7/30	Mile 70, camp by White River. (See page 173.) (Map D)
51	7/31	Mile 78, camp on stream below Griffin Glacier. (Map D)
52	8/1	Mile 88, camp by White River. (Map D)
53	8/2	Mile 98, camp near Russell Glacier. (Map D)
54	8/3	Mile 108, camp below Russell Glacier. (Map D)
55	8/4	Mile 116, camp on flat by Frederika Glacier. (Map D)
56	8/5	Mile 125, camp by ice-filled lake dammed by Nizina Glacier. (See page 179.) (Map D)
57	8/6	Mile 136, camp on Nizina River below glacier. (Map D)
58	8/7–9	Mile 144, camp three nights on spruce flat, Nizina River, building boat. (See page 181.) (Map D)
59	8/10	Camp on Nizina River (near Five-Mile Gulch). (Maps D, E)
60	8/11	Camp on Chitina River near sharp bend shown on Lt. Allen's map. (Map E)
61	8/12–15	Camp four nights at Taral Village on Copper River about three miles below Chitina River. (Map E)
62	8/16	Camp on Copper River at foot of steep bank (near Heney Glacier). (Map E)
63	8/17	Copper River, camp on moraine of Miles Glacier. (Map E)
64	8/18	Camp on Mud flat, Copper River Delta. (See page 200.) (Map E)

CAMP	DATE	LOCATION NOTES
65	8/19–21	Alaganik (stay three nights at fishery with Mike Duval). (Map E)
66	8/22	Mouth of Eyak River, sleep in Columbia River boat. (Map E)
67	8/23	Arrive at Eyak, walk tram to cannery, board the *Salmo*, stay aboard overnight en route to Nuchek. Arrive at Nuchek 1 A.M. on August 24, 1891. (Map E)

Bibliography

Allen, Lieutenant Henry T.
    1886   *Report of an expedition to the Copper, Tanana and Koyukuk Rivers in Alaska.* Vol. 2. Senate Executive Document 125, 49th Congress, 2nd Session, Washington, DC: United States Printing Office.

Association of West Point Graduates
    1893   "Frederick Schwatka. No. 2389, Class of 1871." In *Annual Report, Assoc. of Graduates, West Point, NY*, pp. 67–70. West Point, NY: Association of Graduates.

Baird, Donald
    1965   "Tlingit treasures: How an important collection came to Princeton." *Princeton Alumni Weekly*, Vol. 65, No. 17 pp. 6–11, 17.

Brockman, C. Frank
    1959   *Recreational use of wild lands.* New York Toronto London: McGraw-Hill Book Company.

Brooks, Alfred Hulse
    1916   "Memorial of Charles Willard Hayes." *Bulletin of the Geological Society of America*, Vol. 28, March 31 pp. 80–123.
    1973   *Blazing Alaska's Trails.* 2 ed., Edited by Burton L. Fryxell, with a foreward by John C. Reed, Fairbanks, Alaska: University of Alaska Press.

Coutts, R. C.
    1980   *Yukon: Places & Names.* Sidney, B.C.: Gray's Publishing, Limited.

Cummings, Clara
  1892   "Cryptogams collected by Dr. C. Willard Hayes
                         in Alaska, 1891." *The National Geographic Maga-*
                         zine, May 15 pp. 160–162.
Dall, William H.
  1870   *Alaska and its resources.* Boston: Lee and Shepard.
Dawson, George M.
  1987   *Report on an exploration in the Yukon District,*
                         N.W.T. and adjacent northern portion of British
                         Columbia 1887. Reprinted with permission from
                         Geological and Natural History Survey of Canada,
                         Annual Report (New Series), Vol. III, Part 1 1887–
                         88. Montreal: William Foster Brown & Co. 1889.
                         Whitehorse, YT: The Yukon Historical and Muse-
                         ums Association.
DeArmond, R. N.
  1967   *The founding of Juneau.* Juneau, AK: Gastineau
                         Channel Centennial Association.
Dobrowolsky, Helen
  1988   *Fort Selkirk Bibliography.* Whitehorse, YT: Yukon
                         Tourism Heritage Branch.
Edwards, E. J.
  1899   "Robert Bonner." *Review of Reviews*, August, pp.
                         161–165.
Emmons, George Thornton
  1991   *The Tlingit Indians.* Edited with additions by
                         Frederica de Laguna and a biography by Jean Low.
                         Anthropological Papers of the American Museum
                         of National History, #70., Seattle: University of
                         Washington Press.
Finerty, John F.
  1961   *War-path and bivouac or the conquest of the Sioux.*
                         new ed., second printing, 1962 ed., Norman, OK:
                         University of Oklahoma Press.
Gilder, William H.
  1966   *Schwatka's search. Sledging in the Arctic in quest of*
                         the Franklin Records. New York: Charles Scribner's
                         Sons (reprinted) 1966, New York: Abercrombie
                         & Fitch, Ann Arbor: University Microfilms Inc.
Hayes, Charles W.
  1892   "An expedition through the Yukon District." *The*

	National Geographic Magazine, May 15, pp. 117–159 + pl. 18–20.
1897	"The Yukon District." *Journal of the School of Geography* , Vol. 1, pp. 269–274.
1898	"Copper River as a route to the Yukon basin." *American Geographical Society Bulletin*, Vol. 30, pp. 126–134.

Hayes, Charles W. and Alfred H. Brooks
1900   "Ice cliffs on White River, Yukon Territory." *The National Geographic Magazine*, May pp. 199–201.

Hayes, Charles W. and Sidney Paige
1921   *Handbook for field geologists.* New York: John Wiley & Sons, Inc. London: Chapman & Hall, Limited.

Hebberd, Mary Hardgrove
1952   "Notes on Dr. David Franklin Powell, Known As "White Beaver." *Wisconsin Magazine of History*, Vol 35, No. 4, Summer pp. 306–309.

Heilman, Elizabeth Wiltbank
1935   "Scidmore, Eliza Ruhamah." In *Dictionary of American Biography*, Vol. 8, Part 2. ed. Dumas Malone. pp. 484–485. New York: Charles Scribner's Sons.

Hodgkin, Frank E. and J. J. Galvin
1882   "Frederick Schwatka, U.S.A." In *Pen pictures of representative men of Oregon.* pp. 65–66. Portland OR: Farmer and Dairyman Publ. House.

Hultén, Eric
1968   *Flora of Alaska and neighboring territories.* Stanford, CA: Stanford University Press.

Johnson, Robert E.
1968   "Frederick Schwatka—Cavalry officer, Explorer, Physician." *New England Journal of Medicine* , Vol. 278, pp. 31–35.

Johnson, Robert E., Margaret H. Johnson, Harriet S. Jeanes, and Susan M. Deaver
1984   *Schwatka. The life of Frederick Schwatka (1849–1892), M.D., Arctic explorer, cavalry officer: a Précis.* Montpelier, VT: Horn of the Moon Enterprises.

Klutschak, Heinrich
1987   *Overland to Starvation Cove.* Translated and edited by William Barr. Toronto, Buffalo, London: University of Toronto Press.

Lang, William L.
1983   "At the greatest personal peril to the Photographer."
                Montana The Magazine of Western History, Winter
                pp. 14–28.

Lee, James Melvin
1957   "Bonner, Robert." In *Dictionary of American Biog-
                raphy*, Vol. 1, Part 2. ed. Allen Johnson. pp. 437–
                438. New York: Charles Scribner's and Sons.

McClellan, Catharine
1975   *My Old People Say. An Ethnographic Survey of South-
                ern Yukon Territory. Parts 1 and 2.* National Muse-
                ums of Man Publications in Ethnology, No. 6 (1–
                2), Ottawa: National Museums of Canada.

Merrill, George P.
1960   "Hayes, Charles Willard." In *Dictionary of Ameri-
                can Biography*, Vol. 4, Part 2. ed. Allen Johnson
                and Dumas Malone. pp. 444–445. New York:
                Charles Scribner's and Sons.

Mitcham, Allison
1993   *Taku. The heart of north America's last great wil-
                derness.* Hantsport, Nova Scotia: Lancelot Press.

Monaghan, Frank
1935   "Schwatka, Frederick." In *Dictionary of American
                Biography*, Vol. 8, Part 2. ed. Dumas Malone. pp.
                481–482. New York: Charles Scribner's Sons.

Moser, Jefferson F.
1899   *The salmon fisheries of Alaska: Report of operations
                of the United States Fish Commission steamer Alba-
                tross for the year ending June 30, 1898.* Washing-
                ton, DC: United States Printing Office.

Orth, Donald J.
1967   *Dictionary of Alaska place names.* Geological Sur-
                vey Professional Paper, #567, Washington, D.C.:
                United States Printing Office.

Preble, Edward
1932   "Keely." In *Dictionary of American Biography*, Vol.
                5, Part 2. ed. Dumas Malone. pp.280. New York:
                Charles Scribner's Sons.

Schwatka, Frederick
1884   "The Alaska military reconnaisance of 1883." *Sci-
                ence* , Vol. 3, pp. 220–227, 246–252.

1885a   *Along Alaska's great river. A popular account of the travels of the Alaska Exploring Expedition of 1883, along the great Yukon River, from its soruce to its mouth, in the British Northwest Territory, and in the Territory of Alaska.* New York: Cassell & Co., Ltd.

1885b   "The great river of Alaska." *The Century Magazine*, Sept. pp. 739–751; 819–829.

1885c   *Report of a military reconnaissance in Alaska, made in 1883.* Sen. Ex. Doc. 2, 48th Cong., 2nd Sess., Washington DC: United States Printing Office.

1885d   "The resources of Alaska." *Scientific American*, July 4 pp. 7919–7921.

1886a   "Mineral resources of Alaska." *West Shore*, Vol. 12, pp. 66–67.

1886b   "Mountaineering in Alaska." *Alpine Journal*, Vol. 13, November pp. 89–93.

1888   "Alaska and the Inland Passage." *West Shore*, Vol. 14, pp. 155–157.

1890   "The sun dance of the Sioux." *The Century Magazine*, March, pp. 753–759.

1891   "Two expeditions to Mount St. Elias. I. "The Expedition of the New York Times." *The Century Magazine*, April pp. 865–872.

1892a   "Land of the living cliff-dwellers." *The Century Magazine*, pp. 271–276.

1892b   "The New York Ledger Expedition through unknown Alaska and British America." *The New York Ledger*, Mar. 19, 1–3; Mar. 26, 9–10; Apr. 2, 9–10; Apr. 9, 10–12; Apr. 16, 8–10; Apr. 23, 8–10; Apr. 30, 8–10; May 14, 8–10; May 28, 8–9; Jun. 4, 8–9; Jun. 11, 8; Jun. 18, 8–9; Jul. 9, 8–9; Jul. 16, 8–9; Jul. 23, 13; Jul. 30, 13; Aug. 6, 13; Aug. 13, 13.

1894   *A summer in Alaska. A popular account of the travels of an Alaska exploring expedition along the great Yukon River from its source to its mouth, in the British Northwest Territory, and in the Territory of Alaska.* St. Louis, MO: J. W. Henry.

1895   *In the land of cave and cliff dwellers.* ed. Ada B. Schwatka. New York: Cassell Publishing Co.

1965   *The long Arctic search. The narrative of Lieutenant Frederick Schwatka, U.S.A. 1878–1880, seeking the records of the lost Franklin Expedition.* Marine Historical Assoc. Publ. 44, ed. E.A. Stackpole. Mystic, CT: Marine Historical Association Inc.

Scidmore, Eliza R.
1894   "Recent explorations in Alaska." *National Geographic Magazine*, January 31, pp. 173–179.

Scott, W. B. and E. J. Crossman
1973   *Freshwater fishes of Canada.* Fisheries Research Board of Canada Bull. #184. Ottawa: Fisheries Research Board of Canada.

Seton–Karr, Heywood Walter.
1887a   "The alpine regions of Alaska." Reprinted from, *"Proceedings of the Royal Geographical Society and Monthly Record of Geography." (May No.)* London: Wm. Clowes and Sons, Ltd., pp. 1–17+ maps.
1887b   *Shores and alps of Alaska.* London: Sampson, Low, Marston, Searle and Rivington.

Sherwood, Morgan B.
1979   "The mysterious death of Frederick Schwatka." *Alaska Journal*, Summer pp. 74–75.
1992   *Exploration of Alaska, 1865–1900. Reprinted with preface by Terrence Cole.* Fairbanks, AK: Univ. of Alaska Press.

Teslin Women's Institute
1972   *A history of the settlement of Teslin/Teslin Women's Institute.* Teslin, YT: Teslin Women's Institute.

Thomas, Edward Harper
1970   *Chinook A History and Dictionary of the Northwest Coast Trade Jargon.* Portland, OR: Binfords & Mort.

U.S.C.&G.S.
1896   *Report of the superintendent of the U.S. Coast and Geodetic Survey showing the progress of the work during the fiscal year ending with June, 1895.* Washington, DC: United States Printing Office.

U.S.Congress
1882   "Lieut. Frederick Schwatka." 47th. Congress, 1st. Sess. House Rep. No. 933, Washington, D.C.: United States Printing Office.

Viereck, Leslie A. and Elbert L. Little
1986   *Alaska trees and shrubs.* Reprinted with a new pref-
                ace by Leslie A. Viereck, Fairbanks, AK: Univer-
                sity of Alaska Press.
Webb, Melody
1985   *The last frontier.* Albuquerque, NM: University of
                New Mexico Press.
Williams, William
1889   "Climbing Mount St. Elias." *Scribner's Magazine*, 4
                April, pp. 387–403.
Wright, Allen A.
1992   *Prelude to Bonanza: The discovery and exploration
                of the Yukon.* Whitehorse, YT: Studio North Ltd.
Wright, G. Frederick
1887a   "The Muir Glacier." *American Journal of Science 3rd.
                Ser.*, Vol. 33, No. 1 pp. 1–8.

1887b   "Notes on the glaciation of the Pacific coast." *Ameri-
                can Naturalist*, Vol. 21, pp. 251–256.

NEWSPAPER LIST

Alaskan, The, Sitka, AK
America, Chigago, IL
Daily Alaska Dispatch, The, Juneau, AK
Daily Ledger,The, Tacoma, WA
Juneau City Mining Record, Juneau, AK
Kalamazoo Gazette, The, Kalamazoo, MI
Mason City *Express Republican*, Mason City, IA
Mason City Times,The, Mason City, IA
Morning Globe, The, Tacoma, WA
Morning Oregonian, The, Portland, OR.
New York Herald, New York, NY
New York Times, New York, NY
New York World, New York, NY
Rock Island Daily Argus, The, Rock Island, IL
Seattle Post-Intelligencer, The, Seattle, WA
Sun, The, New York, NY

INDEX